Ventilation of Normobaric and Hyperbaric Objects

Ventilation of Normobaric and Hyperbaric Objects

Ryszard Kłos

CRC Press
Taylor & Francis Group
Boca Raton London New York

CRC Press is an imprint of the
Taylor & Francis Group, an informa business

First edition published 2021
by CRC Press
6000 Broken Sound Parkway NW, Suite 300, Boca Raton, FL 33487-2742

and by CRC Press
2 Park Square, Milton Park, Abingdon, Oxon, OX14 4RN

First edition published by CRC Press 2021
CRC Press is an imprint of Taylor & Francis Group, LLC

ISBN: 978-0-367-67523-3 (hbk)
ISBN: 978-0-367-67524-0 (pbk)
ISBN: 978-1-003-13164-9 (ebk)

Typeset in Times
by SPi Global, India

In appreciation of her assistance and perseverance, I dedicate this book to my dear wife Yvonne, who through the years has supported and aided me in my efforts.

Contents

Preface

This scientific monograph is the result of work and research programs in which the author was a participant:

- Evaluation of diving apparatuses – research sponsored by the Polish Navy (1989–1990)
- Legal, technical, and medical problems of saturation dives – research sponsored by Polish Government Central R&D Project purpose: 35 carried out in 1989–1991
- Diving systems for Polish shelf underwater work – research sponsored by State Committee for Research project № 9937794 C/1993 carried out in 1993–1994
- Principles of interrupted and continuous diving complex ventilation during air expositions – research sponsored by State Committee for Research project № 7 T07C 034 19 carried out in 2001–2002
- Detecting and counteracting terrorist underwater threats performed by divers – research sponsored by State Committee for Research project № O R 0000 98 1 carried out in 2010–2012

as well as projects conducted by the author:

- Diving apparatus for special forces – research sponsored by the Polish Navy carried out in 1989–1992 (Kłos, 2000)
- Evaluation of the semi-closed-circuit diving apparatus for deep diving – research sponsored by the Polish Navy carried out in 1990–1992 (Kłos, 2000)
- Operational examinations of diving equipment – research sponsored by the Polish Navy carried out in 1993–1996
- Oxygen diving technology – research sponsored by State Committee for Research project № 148-101/C-T00/96 carried out in 1996–1998 (Kłos, 2000)
- Principles of ventilation in diving complexes during air exposures by means of oxygen inhalators for decompression and treatment – project sponsored by the Armament Policy Department Ministry of Defence carried out in 1998
- New generation of analytical sensors for submarine atmosphere monitoring – Purposeful Project sponsored by Polish Scientific Research Committee Contract № 148 181/C-T00/98 carried out in 1998–2001 (Kłos, 2008)
- The mathematical models of diving apparatus atmosphere ventilation with partial regeneration of the breathing medium – research sponsored by State Committee for Research project № 0T00A 072 18 carried out in 2000–2002 (Kłos, 2007)
- Submarines life support systems – project sponsored by the Armament Policy Department Ministry of Defence carried out in 2003–2004 (Kłos, 2008)
- Submarine rescue and therapeutic containerized hyperbaric set – project № DPZ/4/TM/W/3.5.3.1/2004/WR sponsored by the Armament Policy Department Ministry of Defence carried out in 2004–2008 (Kłos, 2008)

- Chemisorption of carbon dioxide in military applications – project № 148-414/C-T00/2004 sponsored by the State Committee for Research carried out in 2004–2008 (Kłos & Kłos, 2004; Kłos, 2004a; Kłos, 2004b; Kłos, 2008; Kłos, 2009)
- Protection of miners through the use of escape chambers in the event of an appearance of a harmful breathing atmosphere – project sponsored by KGHM Polska Miedź SA in 2007–2008 (Kłos, 2010)
- The method of saturation dives – project № R00-O0014/3 sponsored by the State Committee for Research carried out in 2007–2009 (Kłos, 2014)
- A new generation of breathing simulator – grant № O N504 497734 sponsored by the State Committee for Research carried out in 2007–2010
- Decompression design for combat missions – project № O R00 001 08 sponsored by the State Committee for Research carried out in 2009–2011 (Kłos, 2011; Kłos, 2012)
- Decompression design for MCM dives – project agreement № DOBR/0047/R/ID1/2012/03 sponsored by the State Committee for Research carried out in 2012–2015 (Kłos, 2016)
- Decompression design for MCM/EOD dives II – project agreement № DOB-BIO8/09/01/2016 sponsored by the State Committee for Research – in progress

During the execution of the aforementioned projects, the author proposed a new mathematical model for the process of ventilation of a semi-closed rebreather with constant dosing of the previously prepared breathing gas (Kłos, 2002a; Kłos, 2007; Kłos, 2016). Its validation required making a special simulator of gas exchange in the breathing process (Kłos, 2002b). The device described here, one of its kind, was designed, tested, and implemented by the author. Use of the device made the experimental verification of the proposed model possible (Kłos, 2003a; Kłos, 2003b).

The author adjusted the above model to make it suitable for the process of ventilation in hyperbaric chambers. The validation process required developing a new type of carbon dioxide emission simulator (Kłos, 2004a; Kłos, 2004b). The author designed and made a prototype of such a simulator, which he used for experiments, thus confirming the assumptions he had made (Kłos & Kłos, 2004; Kłos, 2009).

The generalization of the adopted method for the submarine ventilation process was only an obvious consequence of the earlier considerations (Kłos, 2003a). However, the validation process required the author to undertake extensive research on a real object, which confirmed the validity of the modeling method adopted (Kłos, 2004a; Kłos, 2004b; Kłos, 2008; Kłos, 2015). The research on ventilation of the mining excavation constituted the validation of the adopted research approach (Kłos, 2010).

REFERENCES

Kłos R. 2000. *Aparaty Nurkowe z regeneracją czynnika oddechowego.* Poznań: COOPgraf. ISBN 83-909187-2-2.

Kłos R. 2002a. Mathematical modelling of the breathing space ventilation for semi-closed circuit diving apparatus. *Biocybernetics and Biomedical Engineering,* 22, 79–94.

Kłos R. 2002b. Metabolic simulator supports diving apparatus researches. *Sea Technology*, 12, 53–56.

Kłos R. 2003a. A mathematical model of ventilation process of a distressed/disabled submarine. *Polish Maritime Research,* 4, 23–25.

Kłos R. 2003b. Experimental verification of a new mathematical model of ventilation of closed circuit breathing apparatus. *Polish Maritime Research,* 1, 25–30.

Kłos I. & Kłos R. 2004. *Polish Soda Lime in military applications.* Oświęcim: Chemical Company DWORY S.A. ISBN 83-920272-0-5.

Kłos R. 2004a. Regeneration the atmosphere of a disabled submarine. *Sea Technology*, 2, 51–53.

Kłos R. 2004b. Submarine atmosphere regeneration. *Polish Maritime Research*, 1, 27–30.

Kłos R. 2007. *Mathematical modelling of the normobaric and hyperbaric facilities ventilation.* Gdynia: Polish Hyperbaric Medicine and Technology Society, ISBN 978-83-924989-0-2.

Kłos R. 2008. *Systemy podtrzymania życia na okręcie podwodnym.* Gdynia: Polish Hyperbaric Medicine and Technology Society. ISBN 978-83-924989-4-0.

Kłos R. 2009. *Wapno sodowane w zastosowaniach wojskowych.* Gdynia Polish Hyperbaric Medicine and Technology Society. ISBN 978–83–924989–5–7.

Kłos R. 2010. Komorowy system zachowania życia zabezpieczający górników w przypadku powstania atmosfery niezdatnej do oddychania. *Polish Hyperbaric Research,* 4, 71–88.

Kłos R. 2011. *Możliwości doboru dekompresji dla aparatu nurkowego typu CRABE.* Gdynia: Polish Hyperbaric Medicine and Technology Society. ISBN 978-83-924989-4-0.

Kłos R. 2012. *Możliwości doboru ekspozycji tlenowo-nitroksowych dla aparatu nurkowego typu AMPHORA - założenia do nurkowań standardowych i eksperymentalnych.* Gdynia: Polish Hyperbaric Medicine and Technology Society. ISBN 978-83-924989-8-8.

Kłos R. 2014. *Helioksowe nurkowania saturowane - podstawy teoretyczne do prowadzenia nurkowań i szkolenia.* Gdynia: Polish Hyperbaric Medicine and Technology Society. ISBN 978-83-938322-1-7.

Kłos R. 2015. *Katalityczne utlenianie wodoru na okręcie podwodnym.* Gdynia: Polish Hyperbaric Medicine and Technology Society. ISBN 978-83-938-322-3-1.

Kłos R. 2016. *System trymiksowej dekompresji dla aparatu nurkowego typu CRABE.* Gdynia: Polish Hyperbaric Medicine and Technology Society. ISBN 978-83-938322-5-5.

Acknowledgements

Apart from financing, the presented research required access to unique combat equipment. I express my gratitude to the Navy Command of the Republic of Poland and the 3rd Ship Flotilla for the trust that I received. The research included experiments on humans, which required consent of the Scientific Research Ethics Committee. I thank the Military Medical Institute for their support in obtaining such consent. I would like to express my appreciation to the experimental divers, submarine crews, and mining rescuers from the Rudna Mining Plant in Polkowice for their willingness to take the risk of participating in the research and for having faith in my competencies. I would also like to thank the staff and civilian employees of the Polish Navy and Naval Academy for their help in preparing such extensive experiments.

In particular, I would like to express my gratitude to Dr. Kazimierz Szczepański, who reviewed the entire manuscript.

List of Main Symbols and Abbreviations

$a, b, c, A, B, C\ldots$	constants
α	limit probability, or the level of significance
$\forall$	universal quantification (for all...)
$\exists$	existential quantification (there exists…)
∞	Infinity
$\triangleq$	corresponds to or proportional to
$\gg$	considerably greater ($\ll$ means considerably lesser)
const	constant value
δ	relative value
Δ	difference
$\frac{\partial}{\partial x}$	partial derivative with respect to x
exp	exponent
ε	breathing module
ε_k	apparatus design module
ε_p	pressure module
$\int f(x) \cdot dx$	indefinite integral whose derivative is the given function $f(x)$ with the respect to x
$f(x)$	function f of the variable x
H	depth of water
idem	the same
L	level of confidence
ln	natural logarithm
$\lim_{x \to X} f(x)$	limit of the function $f(x)$ at x approaching the limit X
$\dot{m}$	mass flow rate
$\binom{N}{n}$	Newton's symbol; the number of n-combinations from a given set of N elements (N choose n)
n_X	number of moles of substance X
premix	a fixed-content gas mixture stored in supply bottles integrated with a self-contained diving apparatus or delivered via a supply hose
p	pressure
P	probability
p_X	partial pressure of substance X
π	half of the circumference of the elementary radius
R	universal gas constant
t	time or temperature on the Centigrade scale
T	absolute thermodynamic temperature

$\sum$	sum
SCR	semi-closed circuit rebreather
SCUBA	self-contained *UBA*
UBA	underwater breathing apparatus
W	work
V	volume
$\dot{V}$	volumetric flow rate
$\dot{v}$	volumetric rate of oxygen consumption
$x_n(X)$	mole fraction of substance X
$x_v(X)$	volume fraction of substance X

Introduction

The starting goal of the presented work was to find an adequate mathematical model for the semi-closed-circuit breathing apparatus SCR[1] ventilation with constant *premix*[2] dosage. The aim of the next stage was to apply the model to the hyperbaric chambers, submarines, and sealed mining excavation. The presented model combines a micro and macro scale in normobaric and hyperbaric environments. The aims of the work are presented in Table I.1.

At the beginning, the possibility of using analytical models for ventilation of hyperbaric objects was considered. Elaboration of such a model enables us to relate the physical interpretation of the phenomena occurring in the hyperbaric facilities to the mathematical model. No investigations concerning empiric and semi-empiric models were carried out.

The research was financially supported by the state Scientific Research Committee, the Naval Academy, the government, and private companies. The mathematical models discussed should ensure that the phenomena essential for normobaric and hyperbaric facilities are taken into account. The accuracy of all the parameters that influence these phenomena should also be taken into account. The problem of ventilation is discussed in greater detail than in the prevalent literature and does not concern only the removal of gas contaminants.

The problems that have been presented and analyzed are only a part of the most important problems of ventilation. For example, in the case of semi-closed-circuit breathing apparatuses, the ventilation problems are more related to ensuring the proper oxygen partial pressure than to carbon dioxide removal.

The problem of CO_2 removal is related more to the chemical regeneration of the atmosphere (Kłos & Kłos, 2004; Kłos, 2009). From this point of view, the motion of the gas is very important but not essential – more related to time of contact than ventilation process. Removal of CO_2 from the open-circuit apparatuses is essential whereas maintenance of the proper oxygen partial pressure level is of less importance.

To reach the objectives presented in Table I.1, cooperation between leading diving research centers – DR-DC Toronto (former the Defence and Civil Institute of Environmental Medicine – Toronto), Canada and La Cellule Plongée Humaine et Intervention Sous la Mer (CEPHISMER), Tulon, France – was established. The presented monograph summarizes the objectives shown in Table I.1. At the beginning of each chapter a review of the available literature is made. Research reports prepared in leading research diving centers are analyzed. It has been assumed that such research reports are more complete as compared to books or articles, and easier to interpret. This gives the possibility to formulate opinions on:

- equipment used,
- research reliability,
- complete research,

TABLE I.1
The objectives

Subject:	Mathematical modeling of the normobaric and hyperbaric facilities
Cognitive objectives:	1) A review of literature concerning world achievements in the field of research into modeling normobaric and hyperbaric facilities. 2) Foreign consultations regarding achievements in modeling of normobaric and hyperbaric facilities research. 3) Information exchange and its analysis with leading research centers working on breathing apparatuses 4) Research aimed at development and verification of the mathematical ventilation models of the chosen normobaric and hyperbaric objects. 5) Based on the developed ventilation mathematical models of SCR† with constant *premix*‡ dosage, design instructions of that type of apparatus were determined. Research was carried out to confirm the mathematical model developed.
Operational objectives:	1) This stage of work is aimed at elaboration of methods to carry out research on semi-closed breathing circuit apparatus SCR† with constant dosage of *premix.*‡ The results obtained will make it possible to confirm or deny the validity of the assumed mathematical models for breathing space ventilation. 2) Experimental investigation of the ventilating space of SCR† with constant *premix*‡ dosage. This stage of work is aimed at verification of the developed ventilation mathematical models. 3) If possible, generalization of the assumed ventilation models for SCR† with constant *premix*‡ dosage on other type normobaric and hyperbaric facilities, as well as other semi-closed-circuit apparatuses.
Final standard:	Development of reliable* mathematical ventilation models of some normobaric and hyperbaric facilities.
Methodical directions:	Research on ventilation of hyperbaric objects should be based on domestic capabilities. Particularly the existing experimental base of the Diving Gear and Underwater Work Technology of the Naval Academy should be used and developed.

† Semi-closed-circuit rebreather (breathing apparatus)

‡ *Premix* is a fixed-content gas mixture stored in supply bottles integrated with a self-contained diving apparatus or delivered via a supply hose.

* Checked, validated.

- accuracy of experiments,
- analysis of the problem
- laboratory expertise,
- metrological features of the instruments used,
- correctness of the calibration procedures used, etc.

These features make research reports better and more convenient than those included in articles or books. In most cases, textbooks cause controversy in interpretation of experimental results. Often this is related to limitations imposed by the publisher – for example, those related to the volume of the report. Most research reports are secret. Some of them are for internal use only. Often, obtaining a report is not useful because the content is not designated for a wide range of use. In order to ensure security or protect the invention, some other limitations are introduced. The adopted safety measures should take into consideration the risk that the research results can be used by terrorists. Because this kind of research is expensive, a sponsor who orders it will not give permission for unlimited and free dissemination of the research results. For this reason full use of the findings presented in the papers is strictly limited. However, sometimes there are no difficulties in obtaining it, for example within a framework of military cooperation.

The successive chapters deal with research, the theoretical basis of measurements, its technical realization, elaboration, and discussion of the results obtained. They present complete results of the experimental investigations.

The difficulties inherent in research are twofold. First, research is expensive. The costs are comparable with those of aviation research. Therefore, air and diving research centers are interrelated. In many countries, diving studies are financially supported within the frameworks of aviation and space projects. Introducing diverless technology has resulted in reduced financial support for diving research to an indispensable minimum, which in turn has caused regress in diving technology.

Second, a lot of problems arise when divers are required to participate in experiments. It is stated in international conventions that receiving consent for research with subjects is difficult, as the process requires a great deal of paperwork and passing a personal examination conducted by the Commission in which all lead specialist doctors (professors), nurses, religious representatives, and other ethics specialists are present. It is often difficult to convince such a wide team of people to the presented scientific task (especially military tasks). The consent for research should be given by the responsible authorities, for example in Poland by the Committee for Ethics in Scientific Research (Annex 1).

Diving apparatuses as the basic research objects are presented in Chapters 1–5. The hyperbaric chamber is presented in Chapter 6, submarines in Chapters 7–8, and sealed mining excavation in Chapter 9. Conclusions are presented in Chapter 10.

Chapter 2 introduces the mathematical models for ventilation of breathing space in SCR with constant *premix* dosage. The mathematical models presented fulfill the boundary conditions and take into consideration the influence of partial errors of the measured component values[3] on the final value.[4] The problem of breathing space ventilation of the breathing apparatuses is related to the safety considerations of decompression and generally to diving work. For this reason, it is a basic utilitarian feature of SCR with constant *premix* dosage. The other parameters affecting diving safety are the stabilization rate and the flushing effectiveness of the ecologically confined or semi-confined loop in the breathing apparatuses. Chapter 3 describes the use of the mathematical ventilation models for designing the SCR for constant *premix*.

Chapters 4–5 present the experimental results of the research concerning the mathematical model of ventilation for specified SCR with a metering bellows dispenser.

In Chapter 6, the experimental results of the ventilation mathematical modeling approach for SCR are applied for the ventilation of the hyperbaric chamber. It is shown that the presented model may be generalized for two very different hyperbaric facilities. As follows from the above, the assumed mathematical models are fulfilled in the micro[5] and macro scales[6] as well. The mathematical model presented is the analytical model. It can be assumed, with sufficient approximation, that the physical interpretation of the phenomenon is correct.

In Chapters 7 and 8 the model is generalized for ventilation in submarines and sealed mining excavations, both being large normobaric objects.

The monograph includes only part of the experimental material. Some essential problems relating to hyperbaric ventilation investigations have been omitted. Therefore,

1. The only way to describe the problems dealing with the hyperbaric ventilation was to present the relationship between the phenomena occurring in a hyperbaric chamber, sealed submarine, and mining excavation with those of the SCR with constant *premix* dosage. The problems of ventilation of other types of SCR are more widely discussed in the reports ordered by the Research and Development Department Ministry of Defence[7] (Kłos, 2003b; Kłos, 2011; Kłos, 2012; Kłos, 2016). The problems of hyperbaric ventilation are more widely discussed in the report ordered by the Research and Development Department Ministry of Defence.
2. Many essential problems that had been published earlier were omitted in the description of the problems dealing with the elaboration and experimental verification of the mathematical models of breathing loop in SCRs and closed-circuit rebreathers (CCRs) (Kłos, 2003b; Kłos, 2012).
3. The general schedule of the research program for the SCR leading to practical implementation of the apparatus was presented in the book (Kłos, 2000; Kłos, 2011; Kłos, 2012; Kłos, 2016). The implementation activity was carried out according to the presented schedule. The long-term research program was verified and supervised by the Polish Naval Academy. It is a generalization and a summarization of the experiments dealing with the ventilation model of the diving apparatus breathing space. The purpose and effectiveness of the proposed procedure for determining the basic utilitarian features of the breathing apparatuses are proved in practice. The procedure enables evaluation of practical usage features of the diving apparatuses and opens the possibility for special and particular experimental diving planning. These problems will not be presented here.
4. The elements of the diving apparatus and the description of its construction have been presented in many papers by the author (Kłos, 1999; Kłos, 2000). Most of the papers dealing with the above-mentioned problems have been accepted as a result international consultations. The present method of sharing knowledge of diving apparatuses for underwater works can be used to collect, systematize, and analyze construction and development trends. The division of diving apparatuses presented by the author is not perfect but seems to be one of the most comprehensive in the available field studies. The usability has been confirmed in practice during data systematizing and analysis. Analysis of the

diving trends based on the above-mentioned method had received repeated interest in SCRs in Europe and in the world several years before there was a ready market for these diving apparatuses.[8] Simultaneously the introduction of the three division criteria was the innovative solution in preparing the division of the diving apparatuses. The following division criteria have been introduced: the kind of breathing gas, the range of diving depth, and the principle of operation (Kłos, 2000). Most of the criteria published in the literature review are based only on the constructional criteria of the diving apparatuses.

5. The need for use and problems dealing with the use of an artificial breathing gas in special diving apparatuses were described in the publications (Kłos, 1999; Kłos, 2000; Kłos, 2011; Kłos, 2012; Kłos, 2016). It should be noted that it is a detailed division of the breathing gases used in diving. The history of Polish work focused on the special diving apparatuses, and worldwide work focused on artificial breathing gases has been omitted. In short, this history was described in other publications (Kłos, 2003a).
6. The measurement methods of the most important parameters relating to special diving apparatuses are repeatedly presented in the author's other publications (Kłos, 1999; Kłos, 2000; Kłos, 2011; Kłos, 2012; Kłos, 2015; Kłos, 2016), and therefore it has been omitted from this study.
7. The problems connected with the breathing gas preparation for diving are discussed in the author's other publications (Kłos, 1999; Kłos, 2000). The human reaction to oxygen, contaminants, and inert gases in the hyperbaric environment is related to gas partial pressure. Oxygen, the basic component of the breathing mixtures, is toxic in hyperbaric conditions. Oxygen toxicity in a hyperbaric environment causes many problems during diving (Kłos, 2012). In the hyperbaric environment the inert gases – nitrogen, helium, hydrogen, argon, and neon – become toxic. In the normobaric environment, permissible level of contaminants exceeds the toxicity threshold of them in the hyperbaric environment (Kłos, 2017). Therefore, the diving breathing gas used in deep diving must be of the highest purity and prepared in accordance with special procedures. It is evident that in order to evaluate the purity standards, special gas analyzers and procedures are needed. The applied methods should ensure higher accuracy of measurements. Additionally, due to fire hazard, special oxygen cleanliness[9] should be used for breathing mixtures. The methods of preparing the breathing gases with indications resulting from practical experiences[10] have been repeatedly presented in the author's books (Kłos, 1999; Kłos, 2000). The verified methods for the special preparation and operation of the installation for preparation and distribution of artificial breathing gases are also discussed. The recommended[11] consumables with particular respect to fire hazard are also presented. But in this book these problems are omitted (Kłos, 2014).
8. Very important problems concerning the choice of measurement methods, the breathing gas composition, and contamination percentage tested during experimental saturation diving and research of the prototype breathing systems, particularly with respect to the diving apparatuses, are omitted because they've been described in earlier work (Kłos, 1990; Kłos, 2008; Kłos, 2017).

9. The special diving apparatuses are characterized by the breathing loop that is totally or partly closed. As a result, the breathing gas must not only be supplemented with oxygen but also regenerated because of nascent contaminants.[12] The methods of carbon dioxide removal have not been presented here, as they were presented in the literature earlier (Kłos & Kłos, 2004; Kłos, 2004a; Kłos, 2004b; Kłos, 2009). Recognition of that problem is essential, as it affects the protective period of the diving apparatus functioning. The volume of the diving apparatus space is small enough that the run of the processes in the breathing space is very rapid. The composition of breathing gas mixture can vary rapidly. Deviations from the assumed composition for as short as 1 *min* duration may lead to harm and even death of the diver.
10. The technical problems concerning decompression are important when designing diving apparatuses and developing the diving technology and experimental verification carried out with subjects. The above-mentioned problems are closely connected with the modeling of SCR ventilation. The problems were discussed earlier (Kłos, 1999; Kłos, 2000; Kłos, 2011; Kłos, 2012; Kłos, 2016). Further field studies are planned.

The experiments have been carried out by the research group of the Polish Naval Academy, the Ship and Military Medicine Institute, and also groups from other units. The financial support for these scientific investigations by the State Committee for Research proves that the reviewers recognize the problem of hyperbaric ventilation as essential in hyperbaric technique. The mathematical models of the diving apparatuses enable numerical solutions when it comes to problems with the optimum design and with planning of the diving operations. It facilitates safe and automatic ventilation and improves decompression safety. In the future it would enable the preparation of optimum operational diving conditions in real time. The application of such technology in combat would also be very useful. Associating the combat diving model with the mathematical model for correct decompression prediction would enable the maximum use of the combat equipment, simultaneously improving the tactical and technical properties of the equipment and increasing diving safety.

Modern decompression is carried out with the use of electronic equipment working in real time and taking into account an actual diver's work parameters. The phenomena occurring inside the diving apparatus are not taken into account. Even the use of the most popular gas analyzers for breathing gas will not eliminate the necessity to employ the mathematical models of hyperbaric facility ventilation, as is the case with situations where emergency procedures are to be applied when a measurement error of the breathing gas composition occurs. Gas analyzers will be the basic tools used to design diving apparatuses and will be used for effective planning of special diving operations.

NOTES

1 *Semi-closed **C**ircuit **R**ebreather.*
2 *Premix* is a fixed-content gas mixture stored in supply bottles integrated with a self-contained diving apparatus or delivered via a supply hose.

3 Accuracy of the model's dependent values.
4 Accuracy of the model's independent value.
5 The diving apparatus.
6 The hyperbaric chambers.
7 Presently Armament Politics Department Ministry of Defence.
8 Recreational and technical diving.
9 Purity.
10 Exploitation research.
11 Verified.
12 Mainly CO_2.

REFERENCES

Kłos R. 1990. *Metodyka pomiarów składu mieszanin oddechowych w nurkowych kompleksach hiperbarycznych.* PhD diss., The Polish Naval Academy.

Kłos R. 1999. *Nurkowanie z wykorzystaniem nitroksu.* Poznań: KOOPgraf. ISBN 83-909187-1-4.

Kłos R. 2000. *Aparaty Nurkowe z regeneracją czynnika oddechowego.* Poznań: COOPgraf. ISBN 83-909187-2-2.

Kłos R. 2003a. Podział i badania aparatów nurkowych. *Budownictwo Okrętowe*, 1, 15–20.

Kłos R. 2003b. Experimental verification of a new mathematical model of ventilation of closed circuit breathing apparatus. *Polish Maritime Research*, 1, 25–30.

Kłos I. & Kłos R. 2004. *Polish Soda Lime in military applications.* Oświęcim: Chemical Company DWORY S.A. ISBN 83-920272-0-5.

Kłos R. 2004a. Regeneration the atmosphere of a disabled submarine. *Sea Technology*, 12, 51–53.

Kłos R. 2004b. Submarine atmosphere regeneration. *Polish Maritime Research*, 1, 27–30.

Kłos R. 2008. *Systemy podtrzymania życia na okręcie podwodnym.* Gdynia: Polish Hyperbaric Medicine and Technology Society. ISBN 978-83-924989-4-0.

Kłos R. 2009. *Wapno sodowane w zastosowaniach wojskowych.* Gdynia: Polish Hyperbaric Medicine and Technology Society. ISBN 978–83–924989–5–7.

Kłos R. 2011. *Możliwości doboru dekompresji dla aparatu nurkowego typu CRABE.* Gdynia: Polish Hyperbaric Medicine and Technology Society. ISBN 978-83-924989-4-0.

Kłos R. 2012. *Możliwości doboru ekspozycji tlenowo-nitroksowych dla aparatu nurkowego typu AMPHORA - założenia do nurkowań standardowych i eksperymentalnych.* Gdynia Polish Hyperbaric Medicine and Technology Society. ISBN 978-83-924989-8-8.

Kłos R. 2014. *Helioksowe nurkowania saturowane – podstawy teoretyczne do prowadzenia nurkowań i szkolenia.* Gdynia: Polish Hyperbaric Medicine and Technology Society. ISBN 978-83-938322-1-7.

Kłos R. 2015. Measurement system reliability assessment. *Polish Hyperbaric Research.* 2, 31–46. DOI: 10.1515/phr-2015-0009.

Kłos R. 2016. *System trymiksowej dekompresji dla aparatu nurkowego typu CRABE.* Gdynia: Polish Hyperbaric Medicine and Technology Society. ISBN 978-83-938322-5-5.

Kłos R. 2017. Pollutions of the hyperbaric breathing atmosphere. *Scientific Journal of Polish Naval Academy.* 208, 31–44. DOI: 10.5604/0860889X.1237621.

Part I

Diving Apparatuses

Modeling of the diving apparatus ventilation process is the basis for planning the adequate decompression of the diver during his/her exit from the environment with higher pressure. Determining an adequate model of respiratory space ventilation is particularly important for devices with semi-closed circuit of the breathing mix.

The first part describes the research on ventilation of various types of respiratory system construction. The presented deterministic modeling method was later generalized and used to model the ventilation process of other objects, such as submarines, hyperbaric complexes, or mining excavations, which is described in further parts of the monograph.

1 Ventilation of the Construction with a Constant Dosage System

Most of the old research results presented were from experimental investigations of the German construction of the *underwater breathing apparatuses* UBA type FGG III, a *semi-closed-circuit rebreather* SCR using a constant dosage of *premix*[1] gas mixture, prototypes of the Polish SCR-UBA type GAN-87 using a constant dosage of *premix,* and the self-contained version APW-6M SCUBA (Kłos, 2000). Newer scientific investigations concentrated on the French SCUBA[2] with *alternatively closed and semi-closed-circuit* SCR/CR type AMPHORA (Kłos, 2012). Studies are still continuing, and the research is focused on the French SCR-SCUBA type CRABE (Kłos, 2002; Kłos, 2011; Kłos, 2016).

CONSTRUCTION

This chapter presents the principle of work of the investigated apparatuses with constant dosage of *premix.*

The semi-closed-circuit rebreathers with constant breathing medium dosage type SCR FGG III UBA, SCR GAN-87 UBA, and SCR APW-6M SCUBA, which are self-contained diving apparatuses, have almost the same construction. During normal operation SCR-FGG III UBA and SCR GAN-87 UBA work as a hose supply diving apparatus, but their independent parts play the role of escape diving apparatuses. In contrast to them, the SCR APW-6M SCUBA consists only of an independent version.

The scheme and principle of work for a two-bag version is presented in Figure 1.1.

The autonomic[3] version SCUBA of the diving apparatus is crossed and marked as part B in Figure 1.1. Connecting parts A and B with the supply hose C produces the hose supply version of the diving apparatus, where the breathing medium supply stored in the set 15 is an emergency supply.[4]

In the hose version, the breathing medium from the cylinder (17) is passed via the pressure reducer (12) and selector valve (4) metering nozzle (11) inner supply unit into the supply hose C. The fresh breathing medium passes into the inhalation bag (10), the supply hose C, the return valve (2), and the valve selecting mode of supply (3). In the autonomous version of the diving apparatus UBA, the fresh breathing medium passes from the autonomous set of gas cylinders (15) and the manometer pressure control (13) to the reducer (12) through the cutoff valve (14). The breathing medium from the inhalation bag (10) flows through the hose and inhalation valve of the mouthpiece device (9) into the diver's lungs. Breathing bags are made of an

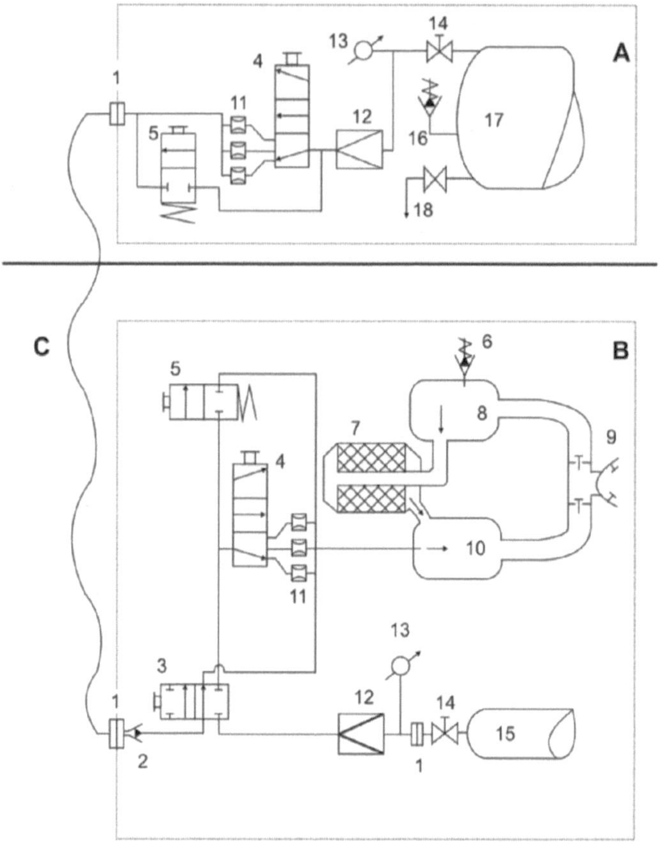

FIGURE 1.1 Two-bag version of the semi-closed-circuit rebreather SCR – underwater breathing apparatus UBA with constant dosage of *premix*. A) External supply unit of the hose version of diving apparatus; B) self-contained version of underwater breathing apparatus SCUBA; C) the supply hose. 1 – coupling, 2 – return valve, 3 – three-pass distribution valve, 4 – distribution valve, 5 – bypass valve, 6 – relief valve, 7 – carbon dioxide scrubber, 8 – exhalation bag, 9 – mouthpiece set with valves and hoses, 10 – inhalation bag, 11 – metering nozzles, 12 – pressure reducer, 13 – manometer, 14 – cutoff valve, 15 – autonomic set of gas cylinders, 16 – safety valve, 17 – external gas stored cylinders, 18 – drainer

elastic, gas-tight fabric, therefore the pressure in the bags approximates the total value of the ambient hydrostatic pressure at the depth of diving and opening overpressure of the relief valve. Exhalation runs through the exhalation valve and the exhalation hose into the exhalation bag (8), then the exhaled gas passes into the carbon dioxide scrubber (7). The relief valve (6), which is mounted on the exhalation bag, is meant to release the excess of exhaled gas into the water. The exhaled breathing medium running through the scrubber gets purified of carbon dioxide and joins the clean breathing medium circuit entering the inhalation bag (10), mixing here with fresh breathing medium. Worthy of note is the value of the reduced breathing medium pressure, which should remain at the level permitting the selected nozzle to work within the range of its supercritical flow maintained for the whole range of diving depths allowed. It means that the mass flow is kept at a constant level within the assumed tolerance.

This type of apparatus is equipped with the special bypass valves (5), designed for filling and washout[5] of the breathing space. In the hose version of the described apparatuses, the bypass valve is externally supplied with gas by a standby diver who remains in contact with the active diver by means of the communication system. Basic tactical and technical data of the diving apparatuses were presented earlier (Kłos, 2007a; Kłos, 2007b).

Only basic tactical and technical data[6] of the SCR FGG III UBA are presented in Table 1.1.

The apparatus was modified for the experimental purposes. The experimental version of this apparatus is the same as that of the type SCR GAN-87 UBA. It is characterized by the same metering values and composition of the breathing gas mixture.

TABLE 1.1
Basic tactical and technical data of the semi-closed-circuit rebreather SCR – underwater diving apparatus UBA type FGG III

Basic tactical and technical data of the diving apparatus type SCR FGG III UBA

Breathing medium	for the depth range:	$29\%_v O_2 + 71\%_v N_2$
	0–60 mH_2O	$17.5\%_v O_2 + 82.5\%_v He$
	50–110 mH_2O	$12.5\%_v O_2 + 87.5\%_v He$
	90–160 mH_2O	$10\%_v O_2 + 90\%_v He$
	140–200 mH_2O	
Working (protection) time	Approx. 3 h (external supply)	
Weight of the apparatus ready to work	in air	280 N
(full bag)	in water	0 N
Dimensions	Width	45.6 cm
	length	68.0 cm
	average height	21.5 cm
Gas cylinders	Number	2
	Material	steel
	Capacity	4 dm^3
	Working pressure	20 MPa
	Experimental pressure	30 MPa
CO_2 absorbent	Type	Soda lime
	granulation	4–6 $mesh$
	Weight of scrubber packing	30 N
Volume of the breathing bag	Inhalation	Approx. 5.5 dm^3
	Exhalation	Approx. 5.5 dm^3
Reduced pressure		Approx. 4 MPa
Metering of the fresh breathing	for the depth range:	$20\ dm^3 \cdot min^{-1}(\pm 10\%)$
medium	0–60 mH_2O	$30\ dm^3 \cdot min^{-1}(\pm 10\%)$
	50–110 mH_2O	$40\ dm^3 \cdot min^{-1}(\pm 10\%)$
	90–160 mH_2O	$50\ dm^3 \cdot min^{-1}(\pm 10\%)$
	140–200 mH_2O	
Pressure relief valve	Type	Plate
	Opening positive gauge pressure	1.1–1.7 kPa

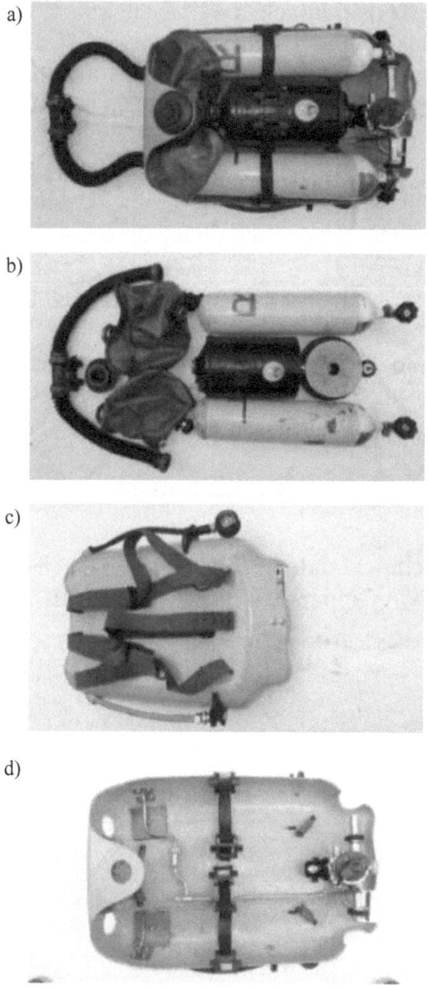

FIGURE 1.2 The basic elements of the SCR FGG III UBA. a) General view of the complex apparatus without the upper cover (head); b) Elements disassembled during normal operation of the apparatus; c) Housing (casing) cover of the diving apparatus – (rear) view from the carrier belts side; d) Housing (casing) cover of the diving apparatus – view from the internal side

More important elements of the apparatus are almost the same as in SCR GAN-87 UBA and SCR APW-6M SCUBA. This is shown in Figure 1.2.

The diving apparatus type SCR GAN-87 UBA is also a semi-closed-circuit rebreather with the constant preproperated breathing medium[7] dosage. Compared with the one-nozzle diving apparatus type SCR APW-6M SCUBA, type SCR GAN-87 UBA is a hose-supply, three-nozzle diving apparatus. The self-contained part works as an escapee diving apparatus. Its principle of operation is the same as the previously discussed construction presented in Figure 1.1.

Basic tactical and technical data are presented in Table 1.2.

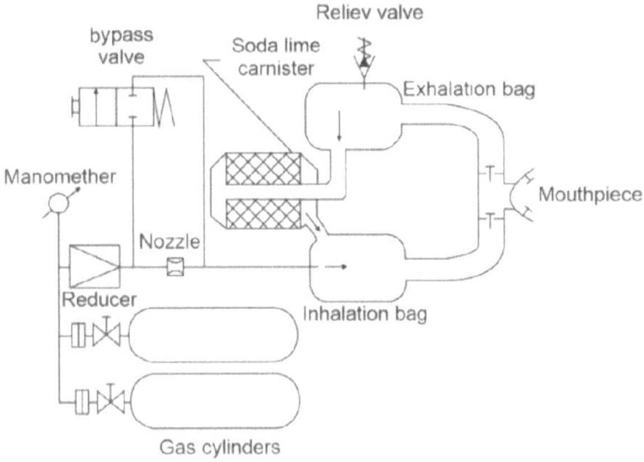

FIGURE 1.3 The diagram of the SCR APW-6M SCUBA

TABLE 1.2
The basic tactical and technical data of the SCR type GAN-87

Basic tactical and technical data of the diving apparatus type SCR GAN-87 UBA

Breathing medium	for the depth range:	$22\%_vO_2/42\%_vN_2/He$
	55–65 mH_2O	$20\%_vO_2/38\%_vN_2/He$
	70–75 mH_2O	$16\%_vO_2/30\%_vN_2/He$
	80–100 mH_2O	$13\%_vO_2/22\%_vN_2/He$
	105–120 mH_2O	
Preparation accuracy	Oxygen	$\pm0.5\%_v$
	Other gases	$\pm1.0\%_v$
Working (protection) time	Approx. 3 *h* (external supply)	
Mass of the apparatus ready to work	in air	Approx. 23 *kg*
(full bag)	in water	Approx. −2 *kg*
Dimensions	Width	45 *cm*
	length	76 *cm*
	average height	21 *cm*
Gas cylinders	Number	2
	Material	steel
	Capacity	3 dm^3
	Working pressure	20 *MPa*
	Experimental pressure	30 *MPa*
CO_2 absorbent	Type	Soda lime
	granulation	4–6 *mesh*
	Volume of scrubber packing	5.4 dm^3
Volume of the breathing bag	Inhalation	Approx. 3.5 dm^3
	Exhalation	Approx. 3.5 dm^3
Reduced pressure		Approx. 2 *MPa*
Metering of the fresh breathing medium	for the depth range:	$(35 \pm 3)\ dm^3 \cdot min^{-1}$
	55–65 mH_2O	
	70–75 mH_2O	
	80–100 mH_2O	
	105–120 mH_2O	
Pressure relief valve	Type	plate
	Opening positive gauge pressure	0.5–1.5 *kPa*

More important elements of the diving apparatus are almost of the same construction as the diving apparatuses type SCR FGG III UBA and SCR APW-6M SCUBA. They differ in the structure of the metering system and the scrubber's dimensions.[8] The purpose behind developing the rebreathers type SCR APW-6M SCUBA and SCR GAN-87 UBA was to achieve unification of the wide range of diving apparatuses.

The diving apparatus type SCR APW-6M SCUBA is a semi-closed-circuit rebreather with constant dosage of *premix*. Basic tactical and technical data are presented in Table 1.3. The diagram of the diving apparatus is shown in Figure 1.3.

The essential elements of the apparatus are listed in Figure 1.4. Its principle of operation is similar to that described already. The difference is that the diving apparatus SCR APW-6M SCUBA is fitted with one nozzle. The diving apparatus features two operating ranges, 0–20 mH_2O and 0–30 mH_2O. The maximum diving depth must be selected before the diving starts, as which of the two standard *Nitrox*[9] gas mixtures will be used depends on this decision. Similarly to the previously discussed constructions, the gas mixture choice is not allowed after the diving commences. The decompression selection rules, detailed presentation of the construction, and the other information have been presented earlier (Kłos, 2000).

TABLE 1.3
Basic tactical and technical data of the SCR type APW-6M

Basic tactical and technical data of the diving apparatus type SCR APW-6M SCUBA

Breathing medium	for the depth range:	
	0–20 mH_2O	$55\%_vO_2 + 45\%_vN_2$
	0–30 mH_2O	$45\%_vO_2 + 55\%_vN_2$
Preparation accuracy	Oxygen	$\pm 0.5\%_v$
	Other gases	$\pm 1.0\%_v$
Working (protection) time	Approx. 2.5 *h*	
Mass of the apparatus ready to work (full bag)	in air	Approx. 23 *kg*
	in water	Approx. −2 *kg*
Dimensions	Width	45 *cm*
	length	76 *cm*
	average height	21 *cm*
Gas cylinders	Number	2
	Material	steel
	Capacity	3 dm^3
	Working pressure	20 *MPa*
	Experimental pressure	30 *MPa*
CO_2 absorbent	Type	Soda lime
	granulation	4–6 *mesh*
	Volume of scrubber packing	3.5 dm^3
Volume of the breathing bag	Inhalation	Approx. 3.5 dm^3
	Exhalation	Approx. 3.5 dm^3
Reduced pressure		Approx. 0.7 *MPa*
Metering of the fresh breathing medium	for the depth range:	$8^{+0,0}_{-0,5}\ dm^3 \cdot min^{-1}$
	0–20 mH_2O	
	0–30 mH_2O	
Pressure relief valve	Type	plate
	Opening positive gauge pressure	0.5–1.5 *kPa*

TABLE 1.4
Derivation of the relationship between oxygen molar fraction *x* and time *t* for the single-bag semi-closed-circuit diving apparatus SCR with the constant metering of *premix* by means of integral calculus

A:	1°	$\dot V \neq f(H)$; $\dot\upsilon \neq f(H)$; $\dot V_E \neq f(H)$	
	2°	$\dot V = \dot V_0 = \dot V(p = p_0)$; $\dot\upsilon = \dot\upsilon_0 = \dot\upsilon(p = p_0)$	
T:		$x(O_2) = f(t)$	
P:	1°	$\frac{p}{R\cdot T}\cdot V\cdot\frac{\partial x}{\partial t} = \frac{p_0}{R\cdot T}\cdot\dot V\cdot x_w - \frac{p_0}{R\cdot T}\cdot\dot\upsilon - \frac{p_0}{R\cdot T}\cdot\left(\dot V - \dot\upsilon\right)\cdot x$	from the oxygen balance in the breathing bag
	2°	$\forall_{a = \frac{p_0}{p}\cdot\frac{\dot V\cdot x_w - \dot\upsilon}{V} = idem;\ b = \frac{p_0}{p}\cdot\frac{\dot V - \dot\upsilon}{V} = idem}$ $\frac{\partial x}{\partial t} = \frac{p_0}{p}\cdot\frac{\dot V\cdot x_w - \dot\upsilon}{V} - \frac{p_0}{p}\cdot\frac{\dot V - \dot\upsilon}{V}\cdot x$	as for the stable circumstances: $\dot V, \dot V_E, \dot v = const$
	3°	$\forall_{a = \frac{p_0}{p}\cdot\frac{\dot V\cdot x_w - \dot\upsilon}{V} = idem;\ b = \frac{p_0}{p}\cdot\frac{\dot V - \dot\upsilon}{V} = idem}$ $\partial x = \left(a - b\cdot x\right)\partial t$	from 2°
	4°	$\frac{\partial x}{b\cdot x - a} + \partial t = 0$	from 3° by dividing by $(b\cdot x - a)$
	5°	$\int\frac{\partial x}{b\cdot x - a} + \int\partial t = C$ where: C is the integral constant	from 4° and integral definition
	6°	$\ln\mid b\cdot x - a\mid \equiv b\cdot(C - t) = C' - b\cdot t$ where-C' –the new constant	From 5°
	7°	$\exp(C' - b\cdot t) = C''\cdot\exp(-b\cdot t) \equiv b\cdot x - a$ where C′ is the new constant	from 6° and natural logarithm definition
	8°	If for boundary conditions $t \to 0 \Rightarrow x \to x_0$, then: $C'' = b\cdot x_0 - a \Rightarrow (b\cdot x_0 - a)\cdot exp(-b\cdot t) = b\cdot x - a$	from7°
	9°	$x(t) = \frac{x_w\cdot\dot V - \dot\upsilon}{\dot V - \dot\upsilon} + \left(x_0 - \frac{x_w\cdot\dot V - \dot\upsilon}{\dot V - \dot\upsilon}\right)\cdot\exp\left(-\frac{p_0}{p}\cdot\frac{\dot V - \dot\upsilon}{V}\cdot t\right)$	from 2° i 8° q.e.d.

A – Assumption, *T* – thesis, *P* – proof (evidence)
Where: $\dot V$ – metering of the fresh breathing medium, $\dot v$ – oxygen consumption, x_w – oxygen molar fraction in the *premix*, $x(t)$ – oxygen molar fraction in the breathing gas mix by diver at the moment t, x_0 – the initial composition of breathing medium in breathing bag ∂t-elementary time, p_0 – normal pressure, p – pressure at the depth of diving, R – universal gas constant, T – absolute temperature, V – breathing bag volume.

VENTILATION

The mathematical models of ventilation in a semi-closed rebreather SCR using constant dosage of *premix* that have been presented in the literature deal with the ventilation of hyperbaric chambers, submarines, and mining excavation (Reimers & Hansen, 1972; Haux, 1982). However, they are not precise enough. Comparison of the presented models with the proposed one will be given in later chapters.

The mathematical models of ventilation for SCR are based on the mass balance of oxygen in a respiratory mixture inhaled by the diver.[10] Mathematical models for SCR are based on the mass balance of oxygen in the respiratory mixture inhaled by the diver.[11] For the time approaching infinity t→∞, the mathematical models proposed here have become the same as the models proposed by the other authors (Reimers &

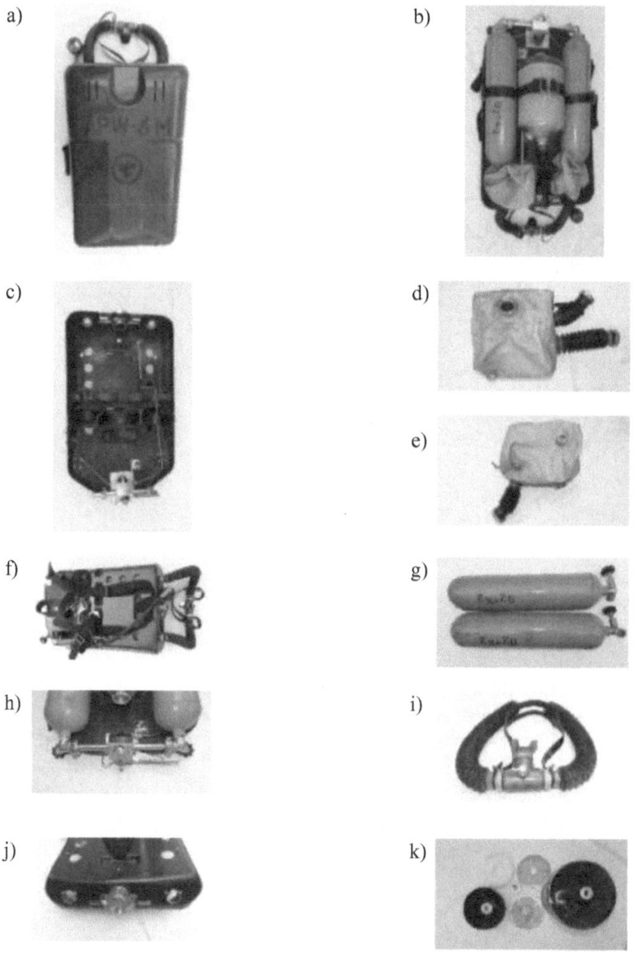

Ventilation

FIGURE 1.4 The basic parts of the SCR type APW-6M. a) General view of the complex diving apparatus; b) General view of the complex diving apparatus without the housing (casing) cover; c) Lower housing (casing) cover; d) Exhalation bag; e) Inhalation bag; f) Harness; g) Supply gas cylinders; h) Reducing-metering assembly; i) Mouthpiece device and the breathing hoses; j) Relief valve; k) Dismantled carbon dioxide scrubber

Hansen, 1972; Williams, 1975; Haux, 1982; Frånberg, 2015). The full-time dependent models are useful during the design of a diving apparatus and suitable decompression. In many cases they are good enough to be used for illustrating the important phenomena occurring in the diving apparatus. Frequently conclusions concerning the composition stabilization rate of the diving apparatus atmosphere after disturbances caused by a change in the oxygen consumption level[12] or caused by periodical ventilation can be drawn based on that model.

The mathematical models for designing SCRs with constant dosage of *premix* that have been presented in the literature ignore the accuracy of the values affecting the composition of the breathing medium. The mathematical model proposed here takes into account the metering accuracy of a breathing medium and the accuracy in calculating oxygen percentage in the *premix*. The limitations of the mathematical models are presented in further chapters.

Classic forms of this type of diving apparatus design can cause errors leading to diving accidents. Diving accidents were recorded during tests performed at medium to high work rate by the diver.

The problem of ventilation is treated here more widely than in most cases, in which the prevailing opinion is that ventilation involves only the removal of contaminants. Despite the comprehensive description of ventilation, the problems presented and analyzed in detail are limited to ensuring proper oxygen partial pressure in the breathing medium inhaled by the diver.

As for the SCR, the ventilation problems are related more to ensuring the oxygen partial pressure than to carbon dioxide removal. The problem of carbon dioxide removal is more related to chemical regeneration of the atmosphere.[13] The problems of carbon dioxide removal from semi-closed-circuit apparatus SCR and experimental results have been previously described (Kłos, 2000; Kłos & Kłos, 2004; Kłos, 2009).

The relative decrease in the oxygen concentration in the inhaled breathing medium compared to the oxygen concentration in the *premix* is observed in SCRs. It is caused by the partly utilized breathing mixture mixed with the *premix* in the bag, schematically presented in Figure 1.5. Oxygen balance presented in Figure 1.5 is valid for the single-bag diving apparatus.

Oxygen is supplied to the breathing bag with the fresh (5) and regenerated breathing medium (4). The oxygen from the breathing bag is removed with the inhaled gas (2) and through the relief valve (1). Part of the oxygen from the inhaled medium is assimilated by the diver's organism (3), and after carbon dioxide removal[14] the breathing medium flows back to the breathing bag (4). Based on the Figure 1.5, the oxygen and breathing medium balance can be calculated. This has been prepared and presented in Tables 1.4–1.6.

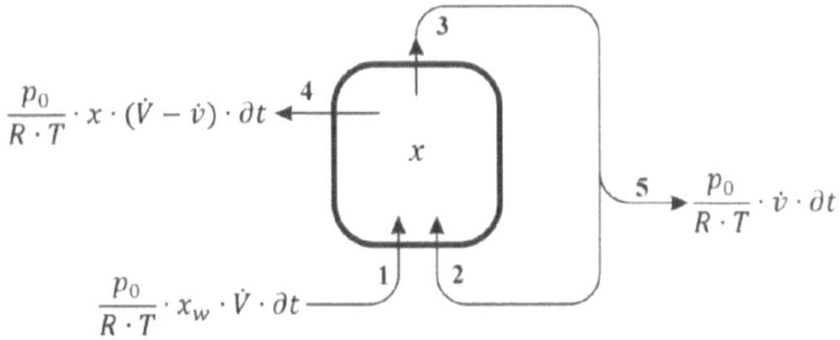

FIGURE 1.5 Oxygen molar balance in the breathing bag, where: $\dot{V}$ – metering of the fresh breathing medium, $\dot{v}$ – oxygen consumption, x_w – oxygen molar fraction in the *premix*, ∂t – elementary time, p_0 – normal pressure, R – universal gas constant, T – the absolute temperature, V – breathing bag volume.

TABLE 1.5
Oxygen molar balance in the breathing bag based on Figure 1.5

		with the fresh breathing medium	with breathing circulation	through the relief valve
Oxygen	Increase	$\frac{p_0}{R\cdot T}\cdot\dot{V}\cdot x_w\cdot dt$	$\frac{p}{R\cdot T}\left[\dot{V}_E\cdot x(i)-\frac{p_0}{p}\cdot\dot{v}\right]dt$	—
	Decrease	—	$\frac{p}{R\cdot T}\cdot\dot{V}_E\cdot x(i)\cdot dt$	$\frac{p_0}{R\cdot T}\cdot\left(\dot{V}-\dot{v}\right)\cdot x(i)\cdot dt$
The whole breathing medium	Increase	$\frac{p_0}{R\cdot T}\cdot\dot{V}\cdot dt$	$\frac{p}{R\cdot T}\cdot\dot{V}_E-\frac{p_0}{R\cdot T}\cdot\dot{v}\cdot dt$	—
	Decrease	—	$\frac{p_0}{R\cdot T}\cdot\dot{V}_E\cdot dt$	$\frac{p_0}{R\cdot T}\cdot\left(\dot{V}-\dot{v}\right)\cdot x(i)\cdot dt$

Where: $\dot{V}$ – metering of the fresh breathing medium, $\dot{v}$ – oxygen consumption, $\dot{V}_E$ – pulmonary ventilation, x_w – oxygen molar fraction in the *premix*, ∂t – elementary time, i – the successive time interval ∂t for substance balance, $x(i)$ – oxygen molar fraction in the breathing bag at the moment $t = i\cdot\partial\tau$, p_0 – normal pressure, R – universal gas constant, T – absolute temperature.

The function describing the relationship between oxygen content in the breathing bag and time is derived from the substance balance presented in Table 1.4 and Table 1.6: $x(O_2) = f(t)$. The relationship $x(O_2) = f(t)$ can be derived on the assumption that premix metering $\dot{V}$, oxygen consumption $\dot{\upsilon}$, and pulmonary ventilation $\dot{V}_E$ are not diving depth dependent H: $\dot{V}\neq f(H)$; $\dot{\upsilon}\neq f(H)$; $\dot{V}_E\neq f(H)$. In reality, a slight dependence on the depth of diving is recorded (Kłos, 2000).

Before equation $x(O_2) = f(t)$ is derived, it is worth noting that the values of premix metering $\dot{V}$, oxygen consumption $\dot{\upsilon}$, and pulmonary ventilation $\dot{V}_E$ are related to normal pressure p_0, therefore the number of oxygen moles should be related to the expanded breathing medium. According to Figure 1.5, Table 1.4, and Table 1.6, the formula describing the oxygen molar fraction in the breathing bag $x(t)$ of SCR can be derived (Kłos, 2007b). The same result was obtained when the problem was being solved by means of the integral calculus and limits calculus. Two conclusions can be drawn from the analysis of equation (9°) from Table 1.4 and Table 1.6:

$$\forall_{t=0}\, x(0) = x_0 \tag{1.1}$$

$$\forall_{t\to\infty}\, x(t) = x_s = \frac{x_w\cdot\dot{V}-\dot{\upsilon}}{\dot{V}-\dot{\upsilon}} \tag{1.2}$$

where: x_s is the stable value of oxygen molar fraction in the inhalation bag $x_s = \lim_{t\to\infty} x(t)$

Equation (1.1) is the simple result of the assumption (8°) from Table 1.4, whereas Equation (1.2) was presented in the literature as the formula obtained from the mass balance of oxygen and breathing medium in the breathing bag (Williams, 1975; Haux, 1982).

TABLE 1.6

The relationship between oxygen molar fraction *x* and time *t* for the single bag semi-closed-circuit diving apparatus SCR with the constant metering of *premix* by means of limits calculus based on oxygen molar balance in the breathing bag presented in Table 1.5.

A:	1°	$\dot{V} \neq f(H);\ \dot{\upsilon} \neq f(H);\ \dot{V}_E \neq f(H)$	
	2°	$\dot{V} = \dot{V}_0 = \dot{V}(p = p_0);\ \dot{\upsilon} = \dot{\upsilon}_0 = \dot{\upsilon}(p = p_0)$	
T:		$x(O_2) = f(t)$	
P:	1°	For elementary time $d\tau$ at time$(i+1)\cdot d\tau$ it can be written: $n(i+1) = \left(n(i) + \frac{p_0}{R\cdot T}\cdot\left[x_w\cdot\dot{V} + x(i)\cdot\dot{V} - \dot{\upsilon} + x(i)\cdot\dot{\upsilon}\right]\partial t\right.$	from oxygen balance
	2°	$x(i+1) = \left[x(i) - \frac{x_w\cdot\dot{V} - \dot{v}}{\dot{V} - \dot{v}}\right]\cdot\left(1 - \frac{p_0}{p}\cdot\frac{\dot{V} - \dot{\upsilon}}{V}\cdot\partial t\right) + \frac{x_w\cdot\dot{V} - \dot{v}}{\dot{V} - \dot{v}}$	from 1° divided by: $n = \frac{p\cdot V}{R\cdot T}$ and $x(i) = \frac{n(i)}{n}$
	3°	$x(i+2) = \left[x(i+1) - \frac{x_w\cdot\dot{V} - \dot{v}}{\dot{V} - \dot{v}}\right]\cdot\left(1 - \frac{p_0}{p}\cdot\frac{\dot{V} - \dot{\upsilon}}{V}\cdot\partial t\right) + \frac{x_w\cdot\dot{V} - \dot{v}}{\dot{V} - \dot{v}}$	from 2°
	4°	$x(i+2) = \left[x(i) - \frac{x_w\cdot\dot{V} - \dot{v}}{\dot{V} - \dot{v}}\right]\cdot\left(1 - \frac{p_0}{p}\cdot\frac{\dot{V} - \dot{\upsilon}}{V}\cdot\partial t\right)^2 + \frac{x_w\cdot\dot{V} - \dot{v}}{\dot{V} - \dot{v}}$	from 3° and 2°
	5°	$x(i+j) = \left[x(i) - \frac{x_w\cdot\dot{V} - \dot{v}}{\dot{V} - \dot{v}}\right]\cdot\left(1 - \frac{p_0}{p}\cdot\frac{\dot{V} - \dot{\upsilon}}{V}\cdot\partial t\right)^j + \frac{x_w\cdot\dot{V} - \dot{v}}{\dot{V} - \dot{v}}$	form 2° ÷ 4°
	6°	For $i = 0, j = t$ and $x(0) = x_0$ it can be written: $x(t) = \frac{x_w\cdot\dot{V} - \dot{v}}{\dot{V} - \dot{v}} + \left[x_0 - \frac{x_w\cdot\dot{V} - \dot{v}}{\dot{V} - \dot{v}}\right]\cdot\lim_{j\to\infty}\left(1 - \frac{p_0}{p}\cdot\frac{\dot{V} - \dot{\upsilon}}{V}\cdot\partial t\right)^j$	from 5°
	7°	$x(t) = a + b\cdot\lim_{j\to\infty}\left[\left(1 + \frac{c}{j}\right)^{j/c}\right]^c$ where: $a = \frac{x_w\cdot\dot{V} - \dot{v}}{\dot{V} - \dot{v}}, b = x_0 - a, c = -\frac{p_0}{p}\cdot\frac{\dot{V} - \dot{\upsilon}}{V}\cdot j\cdot\partial t$	from 6° $\lim_{j\to\infty}(j\cdot dt) = t$
	8°	$x(t) = exp\ (c) + a$, because: $\forall_{a(n)\neq 0}\ \lim_{n\to\infty} a(n) = 0 \Rightarrow \lim_{n\to\infty}\left(1 + a(n)\right]^{1/a(n)} \equiv e$	from 7° $a(n) = \frac{c}{j}$
	9°	$x(t) = \frac{x_w\cdot\dot{V} - \dot{\upsilon}}{\dot{V} - \dot{\upsilon}} + \left(x_0 - \frac{x_w\cdot\dot{V} - \dot{\upsilon}}{\dot{V} - \dot{\upsilon}}\right)\cdot exp\left(-\frac{p_0}{p}\cdot\frac{\dot{V} - \dot{\upsilon}}{V}\cdot t\right)$	from 7°÷8° q.e.d.

A – Assumption, *T* – thesis, *P* – proof (evidence)
Where: $\dot{V}$ – metering of the fresh breathing medium, $\dot{v}$ – oxygen consumption, $\dot{V}_E$ – pulmonary ventilation, x_w – oxygen molar fraction in the *premix*, x_0 – initial composition of breathing medium in breathing bag, ∂t – elementary time, i – the successive time interval ∂t for substance balance, $x(i)$ – oxygen molar fraction in the breathing bag at the moment $t = i\cdot\partial t$, $n(i)$ – number of oxygen moles in the breathing bag at the moment $t = i\cdot\partial t$, p_0 – normal pressure, p – pressure at the depth of diving, R – universal gas constant, T – the absolute temperature, V – breathing bag volume

Equation (1.2) can be used for the two-bag versions of the semi-closed-circuit apparatus with constant breathing medium metering at the normal operation circumstances. It can be proved using the formula for the stable value of oxygen molar fraction in the breathing bag of such apparatus (Table 1.7):

$$\forall_{t\to\infty}\ x(t)=x_s=\frac{x_w\cdot\dot{V}-\dot{\upsilon}}{\dot{V}-\dot{\upsilon}}+\frac{p_0\cdot x_w\cdot\dot{V}}{p\cdot\dot{V}_E} \tag{1.3}$$

For SCR with constant *premix* dosage in the hyperbaric environment: $p_0 < p$, $\dot{V} \ll \dot{V}_E$ and $0 \ll x_w < 1$. It follows that the stable value of oxygen molar fraction in the breathing bag x_s of SCR can be written as Equation (1.2) because $\frac{p_0\cdot x_w\cdot\dot{V}}{p\cdot\dot{V}_E} \ll \frac{x_w\cdot\dot{V}-\dot{\upsilon}}{\dot{V}-\dot{\upsilon}}$.

Stabilization and homogenization of the breathing medium composition is ventilation dependent. It is essential for safety, effectiveness, and economy of diving. First of all, it has a direct bearing on decompression safety. These problems are most important in the design phase of the diving apparatus. An example of the theoretical breathing medium composition in the inhalation/exhalation bag of SCR is presented in Figure 1.6. Composition of gas is the function of the time and oxygen consumed stream during breathing on the surface: $p = p_0$.

In practice, the stabilization time of the breathing medium composition is exceeded by the washout time of the additional breathing space, not the breathing bag, i.e., the inhalation hose, scrubber, etc. Based on Equation (1.2) the formulas enabling calculation of the time necessary for stabilization of the inhale bag atmosphere can be derived.[15] Two criteria for stabilization evaluation can be applied, like in the case of error evaluation. The criterion of absolute stabilization is the first of them: $\Delta x = |x(t) - x(\infty)|$, where Δx represents accepted, limited deviation from the asymptotic value of the oxygen content in the breathing bag. Using equation (9°) from Table 1.4, it can be written: $\Delta x=|\,x(t)-x(\infty)\,|=\left(x_0-\frac{x_w\cdot\dot{V}-\dot{\upsilon}}{\dot{V}-\dot{\upsilon}}\right)\cdot exp\left(-\frac{p_0}{p}\cdot\frac{\dot{V}-\dot{\upsilon}}{V}\cdot t_s\right)$, where t_s represents the time necessary for stabilization of breathing medium composition. The criterion of absolute stabilization is not convenient to compare breathing medium composition stabilization times. For this purpose it is more convenient to apply the relative stabilization criterion δ:

$$\frac{\delta}{100\%}=\frac{|\,x(t)-x(\infty)\,|}{x(\infty)}=\frac{\dot{V}-\dot{\upsilon}}{x_w\cdot\dot{V}-\dot{\upsilon}}\cdot\left(x_0-\frac{x_w\cdot\dot{V}-\dot{\upsilon}}{\dot{V}-\dot{\upsilon}}\right)\cdot exp\left(-\frac{p_0}{p}\cdot\frac{\dot{V}-\dot{\upsilon}}{V}\cdot t_s\right) \tag{1.4}$$

where δ is relative stabilization criterion and t_s is time necessary for stabilization of breathing medium composition.

Based on the analysis of Equations (1.4), it can be stated that the stabilization time is proportional to the pressure that is exerted on the diver: $t_s(p > p_0)=\frac{p}{p_0}\cdot t_s(p_0)$. As it follows, at the lower pressure the stabilization time of breathing medium

TABLE 1.7
Derivation of the formula for stable value of oxygen molar fraction for the two-bag SCR

For: $\dot{V} \gg \dot{v}$

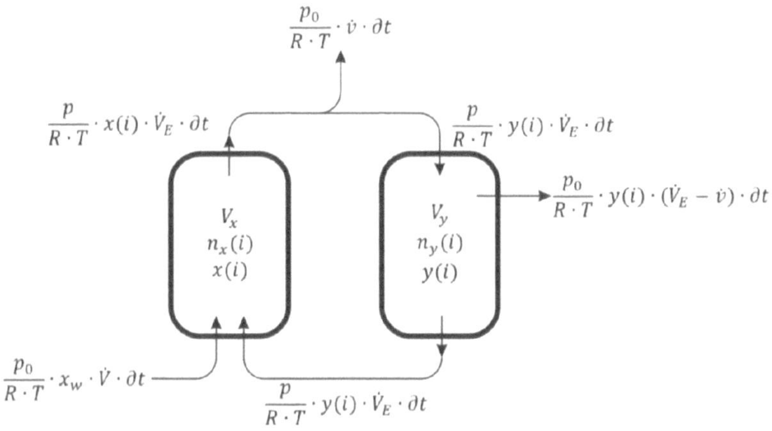

$$n_x(i+1) = n_x(i) + \frac{p_0}{R \cdot T} \cdot \dot{V} \cdot x_w \cdot \partial t + \frac{p}{R \cdot T} \cdot \dot{V}_E \cdot y(i) \cdot \partial t - \frac{p}{R \cdot T} \cdot \dot{V}_E \cdot x(i) \cdot \partial t \,|: n_c(x) = \frac{p \cdot V_x}{R \cdot T}$$

$$\Delta x = \frac{1}{V_x} \cdot \left[\frac{p_0}{p} \cdot \dot{V} \cdot x_w + \dot{V}_E \cdot y(i) - \dot{V}_E \cdot x(i) \right] \cdot \partial t \mid \Delta x = x(i+1) - x(i)$$

For: $\Delta x = 0 \Rightarrow x(i) \to x_s \wedge y(i) \to y_s \quad | \; x_s = lim_{t\to\infty} x(t) \wedge y_s = lim_{t\to\infty} y(t)$

$$\dot{V}_E \cdot (x_s - y_s) = \frac{p_0}{p} \cdot \dot{V} \cdot x_w \quad \to \quad x_s = \frac{p_0}{p} \cdot \frac{\dot{V}}{\dot{V}_E} \cdot x_w + y_s$$

$$n_y(i+1) = n_y(i) + \frac{p}{R \cdot T} \cdot \dot{V}_E \cdot x(i) \cdot \partial t + \frac{p_0}{R \cdot T} \cdot \dot{v} \cdot \partial t - \frac{p_0}{R \cdot T} \cdot (\dot{V} - \dot{v}) \cdot y(i) \cdot \partial t + \frac{p_0}{R \cdot T} \cdot \dot{V}_E \cdot y(i) \cdot \partial t \,|: n_c(x) = \frac{p \cdot V_y}{R \cdot T}$$

$$\Delta y = \frac{1}{V_y} \cdot \left[\dot{V}_E \cdot x(i) - \frac{p_0}{p} \cdot \dot{v} - \frac{p_0}{p} \cdot (\dot{V} - \dot{v}) \cdot y(i) - \dot{V}_E \cdot y(i) \right] \cdot \partial t \mid \Delta y = y(i+1) - y(i)$$

For: $\Delta y = 0 \Rightarrow x(i) \to x_s \wedge y(i) \to y_s \quad | \; x_s = lim_{t\to\infty} x(t) \wedge y_s = lim_{t\to\infty} y(t)$

$$\dot{V}_E \cdot (x_s - y_s) = \frac{p_0}{p} \cdot \left[\dot{v} + (\dot{V} - \dot{v}) \cdot y_s \right]$$

$$y_s = \frac{\dot{V} \cdot x_w - \dot{v}}{\dot{V} - \dot{v}}$$

$$x_s = \frac{p_0}{p} \cdot \frac{\dot{V}}{\dot{V}_E} \cdot x_w + \frac{\dot{V} \cdot x_w - \dot{v}}{\dot{V} - \dot{v}}$$

Where: $\dot{V}$ – metering of the fresh breathing medium, $\dot{v}$ – oxygen consumption, $\dot{V}_E$ – pulmonary ventilation, x_w – oxygen molar fraction in the *premix*, ∂t – elementary time, i – the successive time interval ∂t for substance balance, $x(i)$ – oxygen molar fraction in the inhalation bag at the moment $t = i \cdot \partial\tau$, $y(i)$ – oxygen molar fraction in the exhalation bag at the moment $t = i \cdot \partial\tau$, $n_x(i)$ – the number of oxygen moles in the inhalation bag at the moment $t = i \cdot \partial\tau$, $n_y(i)$ – the number of oxygen moles in the exhalation bag at the moment $t = i \cdot \partial\tau$, $n_c(x)$ – the total mole number in the inhalation bag, $n_c(y)$ – the total mole number in the exhalation bag, p_0 – normal pressure, R – universal gas constant, T – the absolute temperature, V_x – the inhalation bag volume, V_y – the exhalation bag volume.

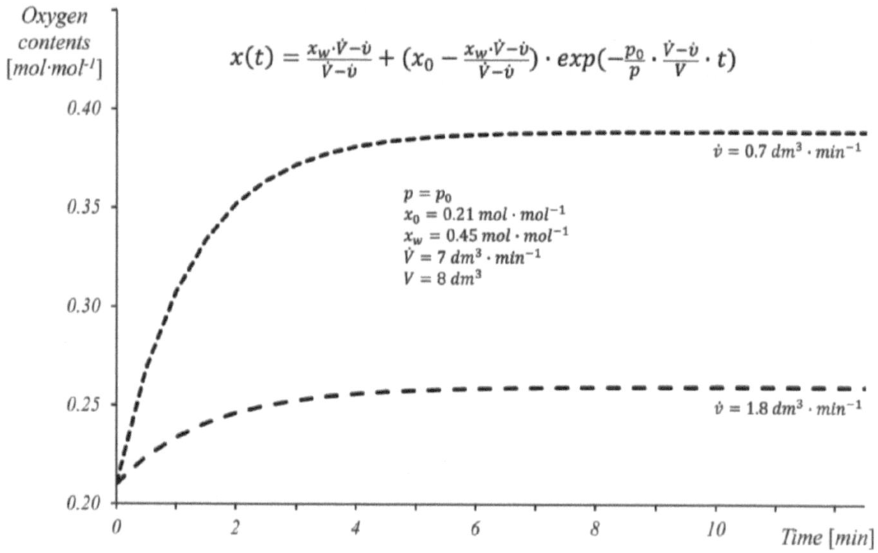

FIGURE 1.6 Theoretical composition of the breathing medium in the breathing bag of a single-bag SCR as the function of time and stream of oxygen consumed during breathing on the surface, where: p – pressure at the depth of diving, p_0 – atmospheric pressure, $\dot{V}$ – metering of the fresh breathing medium, $\dot{v}$ – oxygen consumption, $\dot{V}_E$ – pulmonary ventilation, x_w – oxygen molar fraction in the *premix*, x_0 – the initial composition of the breathing medium in breathing bag.

composition t_s is greater than under higher pressure (Figure 1.7). The diving technology should take into account stabilization time t_s of the breathing medium composition in the breathing space of the diving apparatus.

In cases when a diving bell is used to transport the diver to the work site, up to the certain depth, the diver breathes the air contained in the bell. After exceeding the safety depth at which breathing medium is not hypoxic, the diver can start breathing with a diving apparatus. In order to accelerate the breathing medium composition stabilization, frequent washing of the breathing space is recommended.

Worthy of note is the fact that stabilization of the breathing medium composition in the breathing space is associated with dynamic equilibrium between the fresh and exhaled breathing media. Washing the whole breathing space with a fresh breathing medium does not guarantee immediate equilibrium. However, it will cause the oxygen content in the inhalation bag to approximate the stable value when the diver does not exercise.

As before, the oxygen and breathing medium molar balance is necessary. The balance is presented in Table 1.8. According to this, the balance can be written as follows:

$$x(i+1) = \frac{\frac{p}{p_0} \cdot \left[V_z \cdot x(i)(V_c - V_z) \cdot x_w \right]}{\frac{p}{p_0} \cdot V_c} = \frac{V_z}{V_c} \cdot \left[x(i) - x_w \right] + x_w \tag{1.5}$$

TABLE 1.8
Molar balance of oxygen and the breathing medium in the course of washing of the closed breathing space

		Emptying	Filling
Oxygen	Increase	—	$\frac{p}{p_0}\cdot(V_c - V_z)\cdot x_w$
	Decrease	$\frac{p}{p_0}\cdot(V_c - V_z)\cdot x(i)$	—
	Remainder	$\frac{p}{p_0}\cdot V_z \cdot x(i)$	$\frac{p}{p_0}\cdot\left[V_z \cdot x(i)(V_c - V_z)\cdot x_w\right]$
Breathing medium	Increase	—	$\frac{p}{p_0}\cdot(V_c - V_z)$
	Decrease	$\frac{p}{p_0}\cdot(V_c - V_z)$	—
	Remainder	$\frac{p}{p_0}\cdot V_z$	$\frac{p}{p_0}\cdot V_c$

V_c – the total breathing space volume, V_z – the total residual volume in the breathing space.

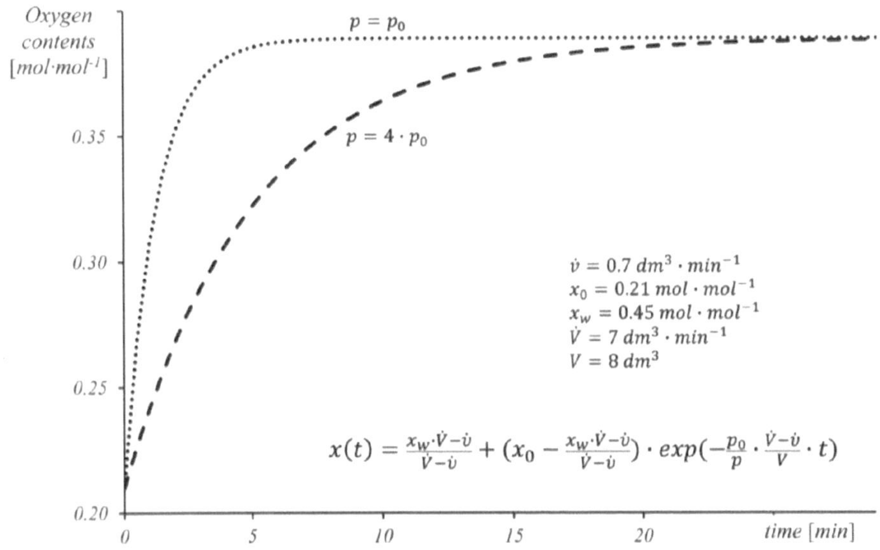

FIGURE 1.7 The theoretical composition of the breathing medium in the breathing bag of SCR as a function of time and pressure at the depth of diving at the same stream of oxygen consumed $\dot{v}$, where: p – pressure at the depth of diving, p_0 – atmospheric pressure, $\dot{V}$ – metering of the fresh breathing medium, $\dot{v}$ – oxygen consumption, $\dot{V}_E$ – pulmonary ventilation, x_w – oxygen molar fraction in the *premix*, x_0 – the initial composition of the breathing medium in the breathing bag.

where V_c is the total breathing space volume and V_z is the total residual volume in the breathing space.[16]

Using Equation (1.5), the formula, which describes the oxygen content in the breathing space after $i + 2$ washing, can be written as follows: $(i+2) = V_z^2 \cdot V_c^{-2} \cdot \left[x(i) - x_w\right] + x_w$. The comparison of this equation with equation (1.5) can be generally written as:

$$x(i,j) = \left(\frac{V_z}{V_c}\right)^j \cdot \left[x(i) - x_w\right] + x_w \tag{1.6}$$

where i is quantity of the moment of time expressed by the basic episode equal to Δt and j is the number of breathing space rinses.

Equation (1.6) enables us to draw an important conclusion that can be written as follows: $x(j) \neq f(H)$, which means that washing of the breathing space is of the same effectiveness for each depth H. The theoretical composition of the breathing medium in the inhalation bag of SCR as a function of time after and without the preliminary washing of the breathing space is shown in Figure 1.8. On assumption that for $i = 0 \Rightarrow x(0) \equiv x_0$, the equation (1.6) can be written as:

$$x(j) = \left(\frac{V_z}{V_c}\right)^j \cdot \left[x_0 - x_w\right] + x_w \tag{1.7}$$

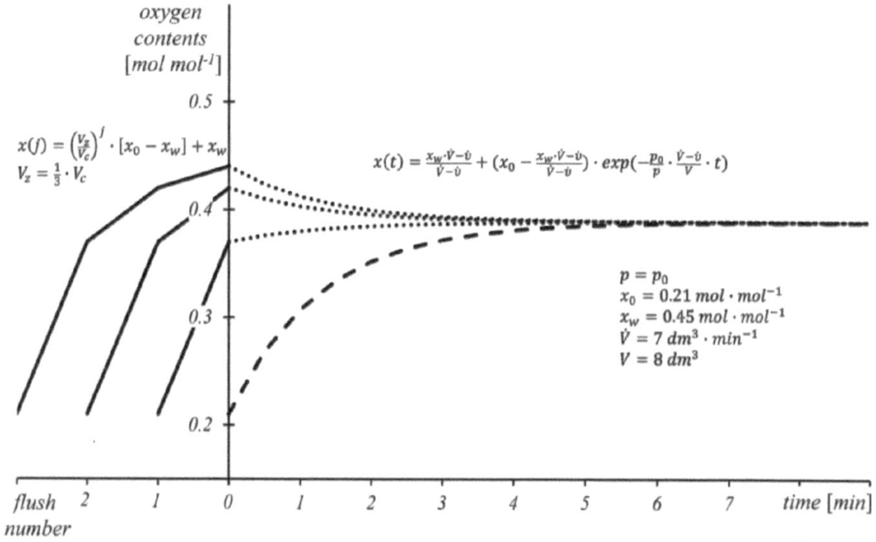

FIGURE 1.8 The theoretical composition of the breathing medium in the breathing bag of SCR as a function of time in the course of breathing on the surface after the washing and without the washing of the breathing space, where: p – pressure at the depth of diving, p_0 – atmospheric pressure, $\dot{V}$ – metering of the fresh breathing medium, $\dot{v}$ – oxygen consumption, $\dot{V}_E$ – pulmonary ventilation, x_w – oxygen molar fraction in the *premix*, x_0 – the initial composition of the breathing medium in the breathing bag.

TABLE 1.9
The results of oxygen content in the inhaled breathing medium for the SCR APW-6M SCUBA

Type of diving apparatus:	SCR APW-6M SCUBA
Breathing medium:	nitrogen-oxygen gas mixture $(29.6 \pm 0.2)\%_n O_2/N_2$
Metering:	$(6.6 \pm 0,1) dm^3 \cdot min^{-1}$
Work load:	sedentary without any movement
Atmospheric pressure	$(101.0 \pm 0.1) kPa$
The temperature:	$(21 \pm 1)\,^\circ C$
Oxygen analyzer:	gas chromatograph (Kłos, 1990)

Time [*min*]	**Oxygen content** [$mol \cdot mol^{-1}$]	**Breathing space volume** [cm^3]	
$\pm 1\ s$	$\pm 0.002\ mol \cdot mol^{-1}$	$\pm 5\ cm^3$	
1	0.239	inhalation bag	3320
5	0.264	exhalation bag	3450
9	0.270	inhalation hose	220
13	0.272	exhalation hose	220
17	0.270	the mouthpiece	45
21	0.270	carbon dioxide scrubber	4070
25	0.270	Total	11325

Preliminary testing of the mathematical model of gas composition stabilization requires experiments in normobaric conditions. The inhaled gas mixture was sampled from the breathing loop of SCR APW-6M SCUBA. The oxygen content in the fresh breathing medium was $x_w = 0.296\ mol \cdot mol^{-2} O_2$. The diver did not work under load except when breathing in the diving apparatus. The experimental results are presented in Table 1.9.

The aim of the experiment was to verify the value of $x(t)$ in the inhaled breathing medium. In order to perform proper recalculations it is necessary to determine the stream of oxygen consumed $\dot{v}$. It can be calculated from the stabilization condition on the assumption that during the experiment the diver was consuming the constant oxygen stream. Using equation (9°) from Table 1.4 it can be written as: $\dot{\upsilon} = \frac{x_w - x(\infty)}{1 - x(\infty)} \cdot \dot{V}$. On the assumption that $x(\infty)$ equals $x(\infty) = 0.270\ mol \cdot mol^{-1}$, the stream of the consumed oxygen $\dot{v}$ can be calculated as follows: $\dot{v} = \frac{0.296 - 0,270}{1 - 0,270} \cdot 6.6 = (0.24 \pm 0.04)\, dm^3 \cdot min^{-1}$. The calculation error was estimated by means of total differential:

$$\Delta\dot{v} = \left|\frac{\partial \dot{v}}{\partial x_w}\right| \cdot \Delta x_w + \left|\frac{\partial \dot{v}}{\partial x(\infty)}\right| \cdot \Delta x(\infty) + \left|\frac{\partial \dot{v}}{\partial \dot{V}}\right| \cdot \Delta\dot{V} \cong 0.04\, dm^3 \cdot min^{-1}$$

$$\left|\frac{\partial \dot{v}}{\partial x_w}\right| \cdot \Delta x_w = \frac{\dot{V}}{1 - x(\infty)} \cdot \Delta x_w \cong \frac{6.6}{1 - 0,270} \cdot 0.002 \cong 0.0181\, dm^3 \cdot min^{-1}$$

TABLE 1.10
The comparison of the oxygen content results in the inhaled breathing medium with the theoretical values determined from the breathing space ventilation mathematical model of the constant *premix* metering SCR.

Time *t*	Measured oxygen content *x*(*t*)	Calculated oxygen content *x*(*t*)
[*min*]	[$mol \cdot mol^{-1}$]	[$mol \cdot mol^{-1}$]
±1 *s*	±0.002 $mol \cdot mol^{-1}$	—
1	0.239	0.236
5	0.264	0.266
9	0.270	0.269
13	0.272	0.269
17	0.270	0.269
21	0.270	0.269
25	0.270	0.269

$$\left|\frac{\partial \dot{v}}{\partial x(\infty)}\right| \cdot \Delta x(\infty) = \frac{1-x_w}{1-x(\infty)} \cdot \dot{V} \cdot \Delta x_w \cong \frac{1-0.296}{1-0.270} \cdot 6.6 \cdot 0.002 \cong 0.0127\, dm^3 \cdot min^{-1}$$

$$\left|\frac{\partial \dot{v}}{\partial \dot{V}}\right| \cdot \Delta \dot{V} = \frac{x_w - x(\infty)}{1-x(\infty)} \cdot \Delta \dot{V} \cong \frac{0.296-0.270}{1-0.270} \cdot 0.4 \cong 0.0142\, dm^3 \cdot min^{-1}$$

The comparison of the measured oxygen content in the inhaled breathing medium and the theoretical value is possible with the use of equation (9°) from Table 1.4 and the measurement data. This comparison is showed in Table 1.10 and Figure 1.9.

As follows from the preliminary experiment, the results showed satisfactory consistence with the mathematical model. The results obtained by means of a gas chromatograph were very accurate, but the number of the measurements was lower because of relatively great single measurement time.[17]

DESIGN OF SCR WITH CONSTANT PREMIX DOSAGE

According to the ventilation model of the single-bag and constant *premix* metering SCR, the function describing the relation between oxygen molar fraction in the breathing gas mixture and time can be presented as equation (9°) from Table 1.4. As time *t* elapses, oxygen molar fraction in the breathing bag *x*(*t*) at the constant stream of the oxygen consumed $\dot{v}$ asymptotically approaches the constant value limit: $\lim_{t\to\infty} x(t) = \frac{x_w \cdot \dot{V} - \dot{\upsilon}}{\dot{V} - \dot{\upsilon}} = x_s = const$. The constant value of oxygen molar fraction in breathing gas mixture x_s continues for several minutes when the gas metering is relatively sufficient. Based on this assumption, the oxygen partial pressure is as follows:

$$\forall_{\dot{v}=const}\; p_s = x_s \cdot p = \frac{x_w \cdot \dot{V} - \dot{\upsilon}}{\dot{V} - \dot{\upsilon}} \cdot p \tag{1.8}$$

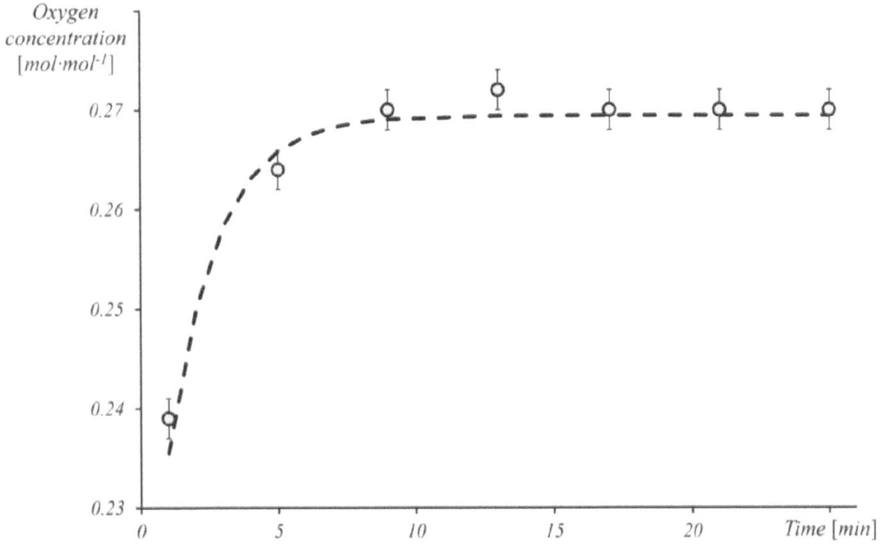

FIGURE 1.9 The comparison of the experimental results of oxygen content in the inhaled breathing medium with the theoretical values determined from the breathing space ventilation mathematical model of the constant *premix* metering SCR.

where p_s is oxygen partial pressure in the breathing gas mixture, p is absolute pressure at the diving depth H, $\dot{V}$ is metering of the *premix*, $\dot{v}$ is oxygen consumption, x_w is oxygen molar fraction in the *premix*, and p is absolute pressure at the diving depth H.

Therefore, it is possible to calculate optimum metering values $\dot{V}$ and oxygen fractions x_w in the *premix* at the assumed value limits of oxygen consumption $\dot{v}$, partial pressure of oxygen p_i, and absolute pressure at the diving depth p. Maximum oxygen partial pressure p_s^{max} occurs when the stream of oxygen consumed is minimum $\dot{v}^{min}$ and the absolute pressure at the diving depth is maximum[18] p^{max}. Minimum oxygen partial pressure p_s^{max} is related to $\dot{v}^{max}$ and p^{min}. Based on that, the following system of equations can be written:

$$\begin{cases} p_s^{max} = \dfrac{\dot{V} \cdot x_w - \dot{v}^{min}}{\dot{V}_0 - \dot{v}^{min}} \cdot p^{max} \\ p_s^{min} = \dfrac{\dot{V} \cdot x_w - \dot{v}^{max}}{\dot{V} - \dot{v}^{max}} \cdot p^{min} \end{cases} \tag{1.9}$$

where p_s^{max} is maximum oxygen partial pressure in the breathing gas mixture, p_s^{min} is minimum oxygen partial pressure in the breathing gas mixture, p^{max} is maximum absolute pressure at the diving depth H, p^{min} is minimum absolute pressure at the diving depth H, $\dot{v}^{min}$ is minimum oxygen consumption, and $\dot{v}^{max}$ is maximum oxygen consumption.

The ranges of values of oxygen partial pressure in the breathing gas mixture p_s, oxygen consumption $\dot{v}$, and absolute pressure at the diving depth p are assumed for the system of Equations (1.9). Values of absolute pressure at the diving depth p follow from the assumed permissible range of diving depths H. Values of the oxygen stream consumed $\dot{v}$ are experimentally determined as a function of the work rate.

TABLE 1.11
Streams of oxygen used and lung ventilation depending on physical effort

Physical effort		Stream of used oxygen	Number of breaths per minute	Lung ventilation	Border stream of used oxygen
Intensity	Example	$[dm^3 \cdot min^{-1}]$	$[min^{-1}]$	$[dm^3 \cdot min^{-1}]$	$[dm^3 \cdot min^{-1}]$
very light	lying on bed	0.25	do 20	8–10	do 0.5
	sitting still	0.30			
	standing still	0.40			
light	walk 3.5 $km \cdot h^{-1}$	0.7	20–25	10–20	0.5–0
moderate	march 6.5 $km \cdot h^{-1}$	2	25–30	20–30	0–5
hard	swimming at speed of 3.0 $km \cdot h^{-1}$	8	30–35	30–50	5–2.0
very hard	running at speed of 13 $km \cdot h^{-1}$	2.0	35–40	50–65	2.0–2.5
extremally hard	running uphill	4.0	>40	>65	>2.5

Table 1.11 presents some estimated values of the oxygen stream consumed $\dot{v}$ that are found in the literature (Przylipiak & Torbus, 1981).

The permissible oxygen partial pressures are experimentally determined and published in the literature. As considerable discussion of this problem was presented earlier (Kłos, 2012), it will not be addressed here. Often, in order to shorten the decompression process, the maximum permissible values of oxygen molar fraction p_s^{max} are applied. The values of *premix* $\dot{V}$ and oxygen molar fraction in *premix* x_w are determined from the equation system (1.9). The scope of the issue relating to the proper choice of optimum gas composition in *premix* x_w and its metering $\dot{V}$ should be broadened. Permissible deviations Δ from the assumed values of *premix* $\dot{V}$ and oxygen molar fraction in *premix* x_w should be taken into account. The reason for introducing the additional parameters is that an accurate preparation of the breathing medium is very troublesome and expensive. Determination of permissible limits of breathing gas metering rates makes it possible to make a decision whether or not the diving apparatus is suitable for use. Taking into consideration the permissible deviations from metering value $\dot{V}$ and oxygen content x_w, the following system of equations can be derived:

$$\begin{cases} p_s^{max} = \dfrac{\left(\dot{V}+\Delta\dot{V}\right)\cdot\left(x_w+\Delta x_w\right)-\dot{v}^{min}}{\dot{V}+\Delta\dot{V}-\dot{v}^{min}}\cdot p^{max} \\ p_s^{min} = \dfrac{\left(\dot{V}-\Delta\dot{V}\right)\cdot\left(x_w-\Delta x_w\right)-\dot{v}^{max}}{\dot{V}-\Delta\dot{V}-\dot{v}^{max}}\cdot p^{min} \end{cases} \tag{1.10}$$

where Δx_w is maximum permissible deviation (upward or downward) from oxygen partial pressure that is fixed as an optimal value in *premix* and $\Delta\dot{V}$ is maximum permissible deviation (upward or downward) from the metering stream of breathing medium that is fixed as an optimal value permissible deviation of *premix* metering.

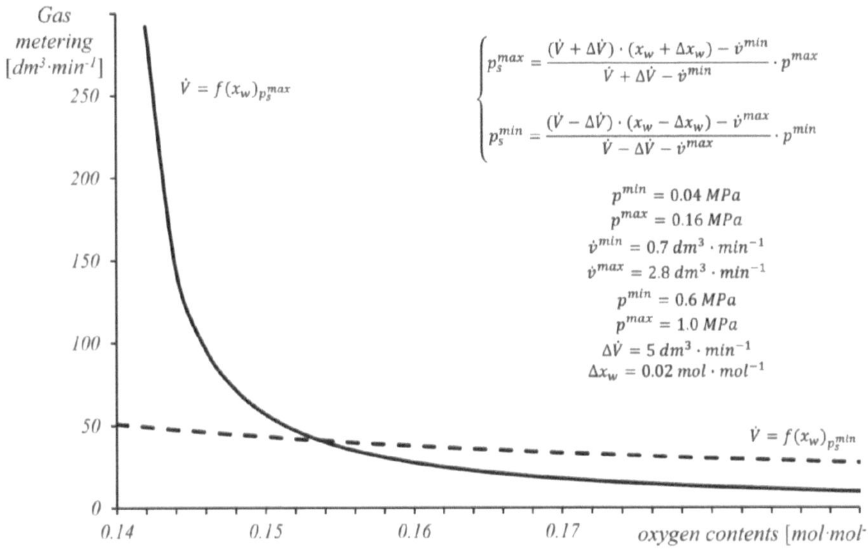

FIGURE 1.10 An example of calculating an optimal value metering of *premix* $\dot{V}$ and oxygen molar fraction in *premix* x_w for an SCR with t *premix* constant metering, where: p_s^{max} – maximum oxygen partial pressure in the breathing gas mixture, p_s^{min} – minimum oxygen partial pressure in the breathing gas mixture, p^{max} – maximum absolute pressure at the diving depth H, p^{min} – minimum absolute pressure at the diving depth H, $\dot{v}^{min}$ – minimum oxygen consumption, $\dot{v}^{max}$ – maximum oxygen consumption, Δx_w – maximum permissible deviation of oxygen contents in the *premix*, $\Delta\dot{V}$ – maximum permissible deviation of the gas metering.

After solving the system of Equations (1.10) with reference to optimal value metering of *premix* $\dot{V}$, a quadratic equation is obtained. The choice of solution to the equation must take into account the physical nature of the phenomenon. The system of Equations (1.10) can be easily solved with the help of computer optimization software. The graphical solution method used to solve the system of Equations (1.10) was presented in Figure 1.10. The relationship between oxygen partial pressure in the inhalation breathing mixture p_s and the stream of oxygen consumed $\dot{v}$ and the depth of diving H calculated on the assumptions is presented in Figure 1.11.

SUMMARY

It is not possible to design an effective and safe diving apparatus that could be adapted for the wide range of the diving depths H and supplied with only one breathing medium having constant oxygen content x_w. A diving apparatus is usually designed for use of several different breathing mediums and metering (Table 1.12). Because there is no one optimal solution for a sufficiently wide range of safe diving depths H,[19] an SCR is usually designed for use of several different *premixes*. Optimum solutions to the design of SCR with constant *premix* dosage[20] are presented in Table 1.12.

In order to obtain the results from the equation system (1.3), the same method was used as in the examples of calculations of optimal values of *premix* $\dot{V}$ and oxygen molar fraction in *premix* x_w, presented in Figures 1.6–1.7. The intersection point of function (1.3) is the optimal solution that must be found.

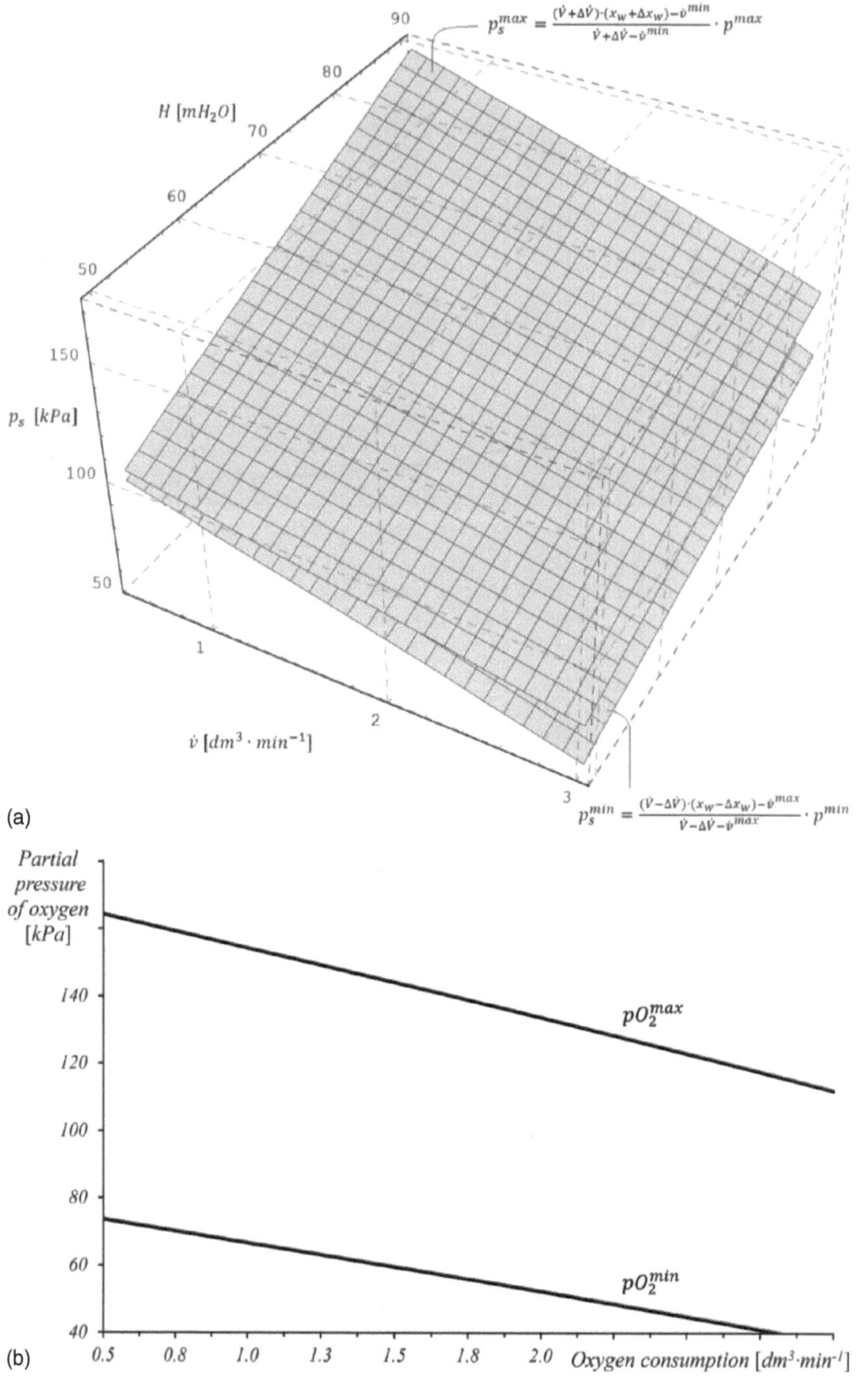

FIGURE 1.11 The relationship between oxygen partial pressure p_s in the breathing mixture and the stream of oxygen consumed $\dot{v}$ and the depth of diving H.

TABLE 1.12

Optimum gas metering and oxygen molar fractions in the prepared breathing medium

Minimum oxygen partial pressure	= 20 kPa
Maximum oxygen partial pressure	= 160 kPa
Minimum oxygen consumption	= 0.7 $dm^3 \cdot min^{-1}$
Maximum oxygen consumption	= 2.8 $dm^3 \cdot min^{-1}$
Metering tolerance	= ± 2.0 $dm^3 \cdot min^{-1}$
Oxygen contents tolerance	= ± 0.02 $mol \cdot mol^{-1}$

H^{max} $[mH_2O]$		H^{min} $[mH_2O]$: 0	10	20	30	40	50	60	70	80	90	100	110
150	x_w				0.099	0.101	0.103	0.104	0.105	0.105	0.106	0.106	0.106
	$\dot{V}$				70.82	54.56	47.59	43.71	41.24	39.53	38.28	37.32	36.56
140	x_w				0.107	0.109	0.111	0.112	0.113	0.113	0.114	0.114	0.114
	$\dot{V}$				58.62	47.29	42.09	39.10	37.17	35.81	34.80	34.02	33.41
130	x_w			0.112	0.116	0.119	0.120	0.121	0.122	0.122	0.123	0.123	0.123
	$\dot{V}$			75.33	49.16	41.15	37.27	34.98	33.46	32.39	31.59	30.96	30.47
120	x_w			0.123	0.127	0.129	0.131	0.132	0.132	0.133	0.133	0.133	0.134
	$\dot{V}$			58.04	41.61	35.91	33.01	31.26	30.08	29.24	28.61	28.11	27.72
110	x_w			0.136	0.140	0.142	0.143	0.144	0.144	0.145	0.145	0.145	
	$\dot{V}$			46.09	35.44	31.37	29.22	27.89	26.99	26.33	25.83	25.44	
100	x_w		0.143	0.151	0.154	0.156	0.157	0.158	0.158	0.159	0.159		
	$\dot{V}$		78.67	37.31	30.31	27.41	25.82	24.82	24.13	23.63	23.25		
90	x_w		0.161	0.168	0.171	0.173	0.174	0.174	0.175	0.175			
	$\dot{V}$		51.40	30.60	25.96	23.91	22.75	22.01	21.50	21.12			

(Continued)

TABLE 1.12 ***(Continued)***

80	x_w		0.183	0.189	0.192	0.193	0.194	0.195	0.195
	$\dot{V}$		36.68	25.29	22.22	20.79	19.97	19.43	19.05
70	x_w		0.209	0.214	0.217	0.218	0.219	0.219	
	$\dot{V}$		27.45	20.98	18.98	18.00	17.42	17.04	
60	x_w		0.242	0.246	0.248	0.249	0.250		
	$\dot{V}$		21.10	17.40	16.12	15.48	15.09		
50	x_w	0.269	0.285	0.288	0.290	0.290	x_w-oxygen contents [$mol \cdot mol^{-1}$]		
	$\dot{V}$	40.05	16.44	14.36	13.58	13.18	$\dot{V}$-*premix* metering [$dm^3 \cdot min^{-1}$]		
40	x_w	0.331	0.342	0.345	0.346				
	$\dot{V}$	20.49	12.86	11.75	11.31				
30	x_w	0.418	0.425	0.427					
	$\dot{V}$	12.74	10.00	9.47					
20	x_w	0.554	0.557						
	$\dot{V}$	8.50	7.63						
10	x_w	0.840							
	$\dot{V}$	3.50							

NOTES

1 *Premix* is a fixed-content gas mixture stored in supply bottles integrated with a self-contained diving apparatus or delivering via a supply hose
2 SCUBA – ***S****elf-****C****ontained* ***U****nderwater* ***B****reathing* ***A****pparatus*
3 Independent.
4 During escape.
5 Additional, forced breathing space ventilation.
6 Provided by the manufacturer.
7 *Premix*.
8 The scrubber construction is the same.
9 Oxygen-nitrogen gas mixtures.
10 For the time approaching infinity $t \to \infty$.
11 In most cases estimated for the time approaching infinity $t \to \infty$.
12 E.g., caused by a change in work rate.
13 Where the gas motion is an important but not essential problem, for example by the contact time.
14 Regeneration.
15 Depending upon the assumed criterion.
16 "Dead space."
17 Approximately 4 *min*.
18 Maximum depth of diving H^{max}.
19 I.e., optimal composition of *premix* x_w and its metering $\dot{V}$.
20 The solution of the system of Equations (1.10).

REFERENCES

Frånberg O. 2015. Oxygen content in semi-closed rebreathing apparatuses for underwater use: Measurements and modeling. PhD diss., Stockholm School of Technology and Health. ISSN 1653-3836.

Haux G. 1982. *Subsea manned engineering*. London: Bailliére Tindall. ISBN 0-7020-0749-8.

Kłos I. and Kłos R. 2004. *Polish Soda Lime in military applications*. Oświęcim: Chemical Company DWORY S.A. ISBN 83-920272-0-5.

Kłos R. 1990. Metodyka pomiarów składu mieszanin oddechowych w nurkowych kompleksach hiperbarycznych. PhD diss., The Polish Naval Academy.

Kłos R. 2000. *Aparaty Nurkowe z regeneracją czynnika oddechowego*. Poznań: COOPgraf. ISBN 83-909187-2-2.

Kłos R. 2002. Mathematical modelling of the breathing space ventilation for semi-closed circuit diving apparatus. *Biocybernetics and Biomedical Engineering* 22, 79–94.

Kłos R. 2007a. Modelowanie procesów wentylancji obiektów normo- i hiperbarycznych. Higher PhD diss., The Polish Naval Academy. PL ISSN 0860-889X nr 160A.

Kłos R. 2007b. *Mathematical modelling of the normobaric and hyperbaric facilities ventilation*. Gdynia: Polish Hyperbaric Medicine and Technology Society. ISBN 978-83-924989-0-2.

Kłos R. 2009. *Wapno sodowane w zastosowaniach wojskowych*. Gdynia: Polish Hyperbaric Medicine and Technology Society. ISBN 978-83-92499889-5-7.

Kłos R. 2011. *Możliwości doboru dekompresji dla aparatu nurkowego typu CRABE*. Gdynia: Polish Hyperbaric Medicine and Technology Society. ISBN 978-83-924989-4-0.

Kłos R. 2012. *Możliwości doboru ekspozycji tlenowo-nitroksowych dla aparatu nurkowego typu AMPHORA - założenia do nurkowań standardowych i eksperymentalnych*. Gdynia: Polish Hyperbaric Medicine and Technology Society. ISBN 978-83-924989-8-8.

Kłos R. 2016. *System trymiksowej dekompresji dla aparatu nurkowego typu CRABE*. Gdynia: Polish Hyperbaric Medicine and Technology Society. ISBN 978-83-938322-5-5.

Przylipiak M & Torbus J. 1981. *Sprzęt i prace nurkowe-poradnik.* Warszawa: Wydawnictwo Ministerstwa Obrony Narodowej. ISBN 83-11-06590-X.

Reimers SD & Hansen OR. 1972. *Environmental control for hiperbaric applications*. Panama City: US Navy Experimental Diving Unit. NEDU Rep. 25–72.

Williams S. 1975. Engineering principles of underwater breathing apparatus. In *The physiology and Medicine of diving*, ed. PB Bennett & DH Elliott, 34–46. London: Baillière Tindall.

2 Unmanned Research on *SCR* with Constant *Premix* Dosage

Implementation of the new diving technologies and constructions developed carried some risk related to manned experiments. Therefore, a laboratory stand was built that could be used to carry out unmanned research on the underwater breathing apparatus UBA. Its use should reduce the risk in such research. Models of a similar laboratory stand prototype have been developed in some countries. The most similar one is used in Sweden (Loncar, 1992; Kłos, 2002; Frånberg, 2015).

Modernization of the laboratory stand *breathing machine* was the basic technical problem that was solved before unmanned research was begun. The aim of the modernization was to develop a pressure attachment that enables oxygen uptake from the breathing atmosphere, then humidifying and replacing oxygen taken from the atmosphere by a suitable amount of carbon dioxide emitted. A solution to this problem was to enable modeling the processes occurring in SCR. This was expected to save time, costs of diving equipment design, and costs of developing decompression schedules. As there was no verified mathematical model available, it was necessary to reconstruct the prototype after each series of the preliminary investigations. Moreover, the verified model was to make it possible to evaluate the diving systems in use.

ASSUMPTIONS

To simulate the respiratory process, a mechanical representation of pulmonary ventilation and gas exchange in the course of breathing is needed. The processes should be precisely and repeatedly represented, although there is no need to have each and every detail of the real respiratory cycle represented.

Usually, it is assumed that an approximated shape of respiration should be sinusoidal (Kłos, 2007a; Kłos, 2007b). When the body gas exchange is simulated, attention should be focused on oxygen consumption and its replacement by the carbon dioxide emitted. A respiratory quotient defines the volumetric ratio for the emitted carbon dioxide that replaces the oxygen consumed. Two values of *respiratory quotient*[1] $\varepsilon_{RQ} = \{1.0;0.8;0.75\}_v$ are most often used.

RESEARCH ON CHEMICAL REACTIONS

The process is simulated by combustion of catalyst chemical substances in a reactor. The respiratory simulator makes volumetric uptake of chemical substances from the breathing space and catalyst combustion possible, then the volume containing the combustion products is recycled to the SCR breathing loop. The following chemical reactions describe the process:

$$
\begin{aligned}
&2\cdot CH_3CHO+5\cdot O_2 \xrightarrow{catalyst} 4\cdot CO_2+4\cdot H_2O \qquad \varepsilon_{RQ}=0.80\\
&\left(CH_3\right)_2 CO+4\cdot O_2 \xrightarrow{catalyst} 3\cdot CO_2+3\cdot H_2O \qquad \varepsilon_{RQ}=0.75
\end{aligned}
\tag{2.1}
$$

Acetone $(CH_3)_2CO$ was used in the research on the catalytic combustion process. To maintain the required rate of acetone a Lubrizol Company pump type $40-PS4L$ was used. The diver's oxygen consumption was assumed to be $\dot{v}=\left(0.3;\,2.4\right)dm^3\cdot min^{-1}$ (Table 1.11). Therefore, acetone metering was maintained within the range of $\dot{u}=\left(0.2;\,1.8\right)cm^3\cdot min^{-1}$.

To evaluate the efficiency of the catalyst, a measuring stand was built. It is presented in Figure 2.1.

The metering pump (5) shown in Figure 2.1 maintains satisfactory accuracy of counterpressure liquid metering. To check the pump characteristics, indirect measurements of the counterpressure app. 20 MPa were carried out (Kłos, 2007b).

The preliminary research on acetone $(CH_3)_2CO$ oxidation was carried out with the use of the catalyst,[2] type 0.3R73 manufactured by *Johnson Matthey Limited.* A glass pipe 70 mm in diameter was filled with 600 g of the catalyst. The catalyst bed in the glass pipe had app. 18.5 cm in height. The air flow rate was maintained at the level of $(2.05 \pm 0.05)m^3 \cdot h^{-1}$ (app. 33 $dm^3 \cdot min^{-1}$). From the glass reactor, the air then passed through a gas chromatograph metering valve coupled with a mass spectrometer GC/MS or a gas analyzer set, type *Multiwarn* (Figure 2.2).

Details concerning the type of operating parameters of gas chromatograph and mass spectrometer GC/MS are presented in Table 2.1.

Multiwarn gas analyzer's operating parameters are given in Table 2.2.

Initially, the composition of gas mixture that left the top part of the catalyst bed was measured at three temperatures of reaction: 350°C, 400°C, and 450°C (Table 2.3).

The *Multiwarn* gas analyzer is equipped with a *pellistor* sensor[3] that enables detecting presence of combustible gases[4] in these conditions. By checking *Multiwarn*'s indications for CO and CH_4 a good correlation[5] for this measurements was found. As it follows from the above and GC/MS measurements, no hydrocarbons were observed (Table 2.4).

Based on these measurements, it can be concluded that the catalyst used in the experiment can be also used for constructing an oxygen uptake attachment, which also enables carbon dioxide and water vapor emission.

An analysis of combustible gases with the *Multiwarn* gas analyzer cannot be carried out in these conditions, as their presence is masked by carbon oxide. The theoretical relationship between the oxygen and carbon dioxide content during a complete combustion can be calculated taking into account the variation of volume. The theoretical relationships concerning liquid water are as follows:

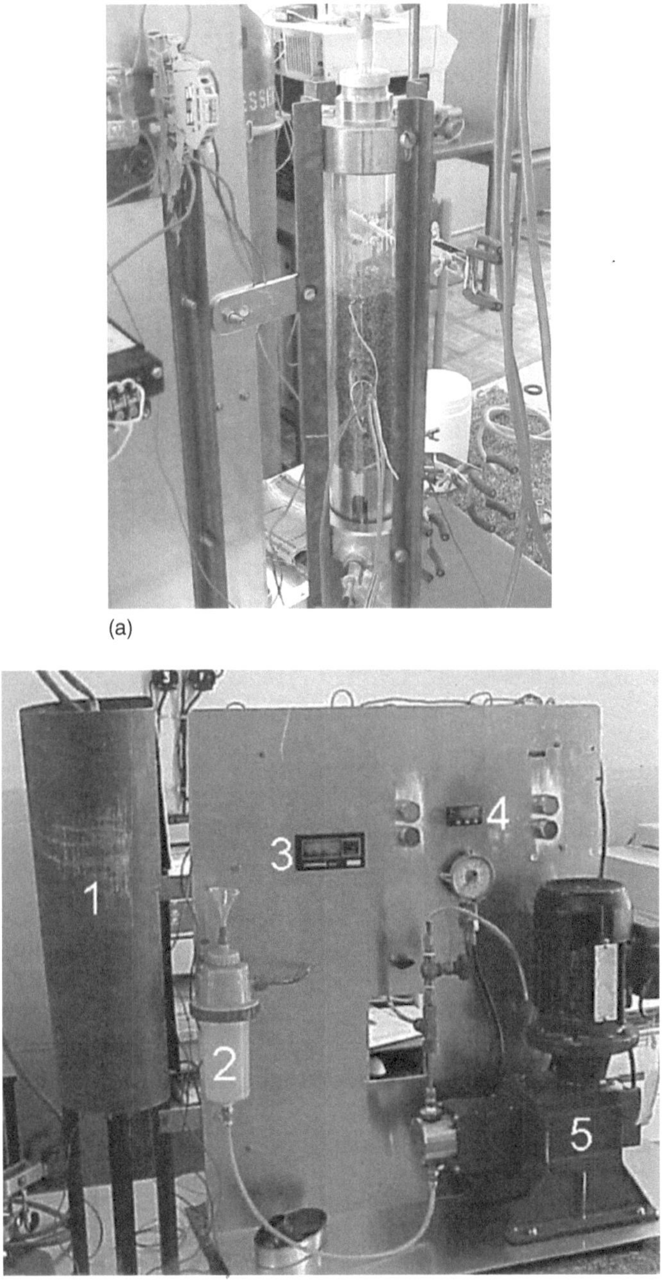

FIGURE 2.1 Laboratory stand for research on catalyst efficiency: a) reactor, b) measuring stand, 1 – reactor in casing (version 1), 2 – reacting substance container, 3 – temperature multipoint measuring set display, 4 – temperature regulator, 5 – metering pump.

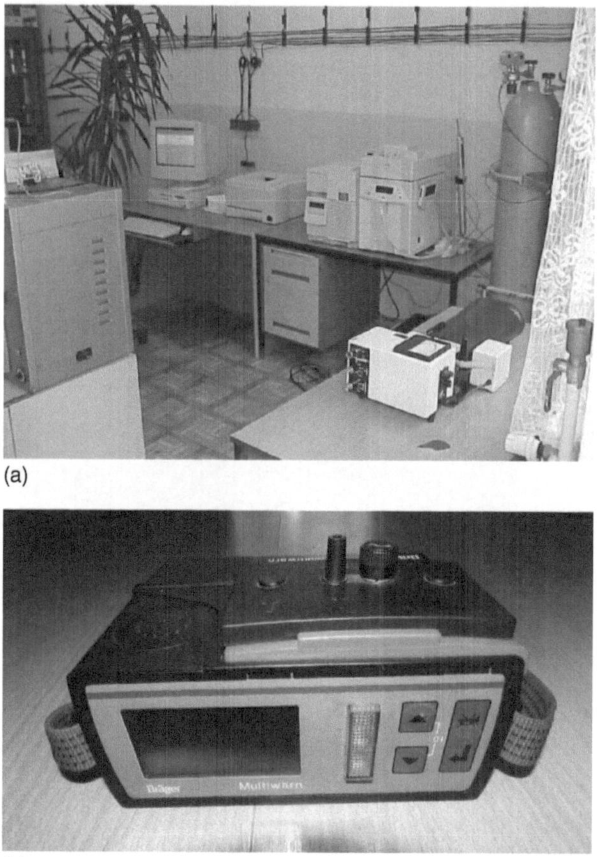

(a)

(b)

FIGURE 2.2 Laboratory stands: a) gas chromatograph coupled with mass spectroscope GC/MS; b) multichannel gas analyzer *Multiwarn* type.

TABLE 2.1
Operating conditions of gas chromatograph and mass spectroscope

Gas chromatograph type AGILENT 6850 GC	
Mass spectroscope type AGILENT 5973 MSD	
capillary chromatographic column:	
type: 190S1S-433E HP-5MS,	
length: 30m / diameter of 250 μm,	
thickness of deposited packaging layer 0.25 μm	
analytical duration	20 min
sample application	manually
carrier gas	helium
flow rate	$1.20 \cdot 10^{-3}\ dm^3 \cdot min^{-1}$
inlet temperature	200°C
the ions source temperature	230°C
quadrupole temperature	150°C

TABLE 2.2
Selected technical data of the *Multiwarn* analyzer (Dräger Safety AG & Co. KGaA, 2003)

Ambient conditions:	–20–40°*C*, 700–1300 *hPa*, 10–95%$_R$			
Time of work:	>8 *h*			
Charging time:	every 3 *weeks*			
Dimensions:	(55 × 110 × 65)*mm*			
Weight:	1 *kg*			
Sensor:	CO_2	CO	CH_4	O_2
Principle of operation:	*IR* spectrophotometr	electromechanical	pellistor	magnetodynamic
Measurement range:	0–25%$_v$	0–200 *ppm*	0–100 % *LEL*†	0–25%$_v$
Zero repeatibility:	≤ ±0.01%$_v$	≤ ±2*ppm*	≤ ±1 % *LEL*†	≤ ±0.2%$_v$
Sensitivity‡:	≤ ±5%	≤ ±1%	≤ ±2.5%	≤ ±1%
Pressure effect*:	≤ ±0.16%	≤ ±0.01%	≤ ±0.1%	≤ ±0.01%
Long-term drift**:	≤ ±0.4%	≤ ±1%	≤ ±1%	≤ ±1%
Chemical factors affecting measurement:	aldehydes, ketones, water	hydrocarbons and other organic compounds	carbon oxide and organic compounds	chlorine, ethane, nitrogen dioxide

† Lower Explosive Limit, ‡ of measured value, * of measured value per 1 *hPa*, ** of measured value per month

$$\left(CH_3\right)_2 CO + 4\cdot O_2 \xrightarrow{catalyst} 3\cdot CO_2 + 3\cdot H_2O \qquad \varepsilon_{RQ} = \frac{V_{CO_2}}{V_{O_2}} = \frac{3}{4}$$

$$\left.\begin{aligned} V_{O_2} &\cong 0.209\cdot V \\ V_{CO_2} &\cong 0.000\cdot V \\ V_{N_2} &\cong 0.780\cdot V \\ V_{residual} &\cong 0.011\cdot V \end{aligned}\right\} \xrightarrow[reaction]{cathalitic} \left\{\begin{aligned} V_{O_2} &\cong 0.209\cdot V - \frac{4}{3}\cdot V_{CO_2} \\ V_{CO_2} &\cong V_{CO_2} \\ V_{N_2} &\cong 0.780\cdot V \\ V_{rest} &\cong 0.011\cdot V \end{aligned}\right.$$

$$Sum: \qquad V \qquad\qquad V - \frac{1}{3}\cdot y$$

where V is volume of reagents, V_{CO_2} is volume of carbon dioxide, and V_{O_2} is volume of oxide;

$$\begin{cases} x_{CO_2} \cong \dfrac{V_{CO_2}}{V - \frac{1}{3}\cdot V_{CO_2}} \quad \Rightarrow \quad V \cong \dfrac{V_{CO_2}}{x_{CO_2}} + \dfrac{1}{3}\cdot V_{CO_2} \\ x_{CO_2} \cong \dfrac{0.209\cdot V - \frac{4}{3}\cdot V_{CO_2}}{V - \frac{1}{3}\cdot V_{CO_2}} \end{cases}$$

TABLE 2.3
Catalyst bed temperatures and measurement results obtained with the gas analyzer type *Multiwarn*.

Catalyst temperature [±2°C]			Components' concentrations during measurements			
at the bottom	middle part [°C]	upper part	O_2 [%$_v$]	CO_2 [%$_v$]	CO [ppm]	CH_4 [%LEL]
±2°C	±2°C	±2°C	[±0.2%$_v$]	[±0.05%$_v$]	[±2 ppm]	[±2 % LEL]
350	515	433	19.44	0.58	376.98	1.43
			12.99	5.22		
			10.72	6.17		
247	360	357	20.90	0.11		
			19.95	0.40	40.37	0.22
			12.85	5.22	891.43	12.06
			13.65	5.81		
237	400	400	19.16	0.94	66.25	0.27
			14.09	4.59		
			14.48	5.06	1000.55	11.65
278	470	400	20.90	0.14		
			20.62	0.22		
			14.57	3.99		
268	422	400	20.90	0.06		
			20.90	0.08		
			19.52	0.70	65.50	0.39
			13.87	4.66		
			16.30	4.05	431.53	3.91
209	360	350	19.57	1.34		
			20.39	0.46		
			17.48	1.67		
			13.12	5.00	500.00	3.72

TABLE 2.4
Example of experimental GC/MS results

(a) MS spectrum of reaction products after catalyst conversion within the temperature range of 250–350°*C*

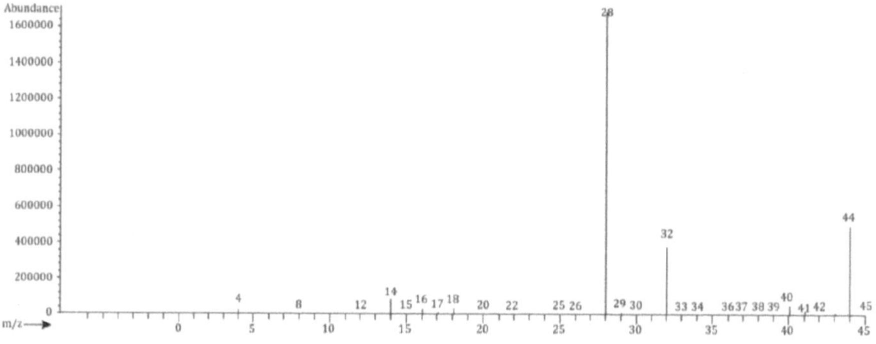

TABLE 2.4 *(Continued)*

(b) MS spectrum of reaction products after catalyst conversion within the temperature range of 350–450°*C*

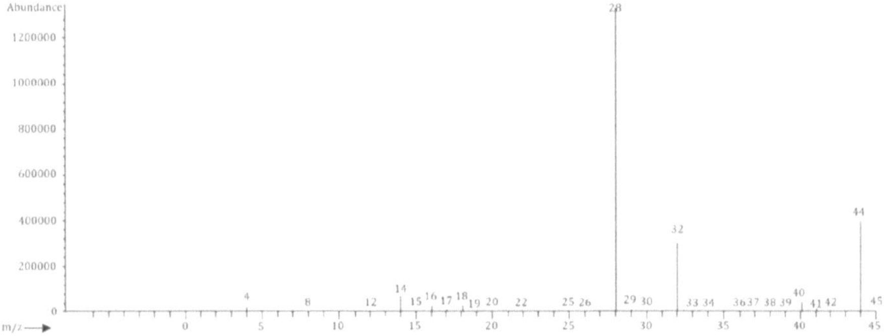

(c) MS spectrum of acetone

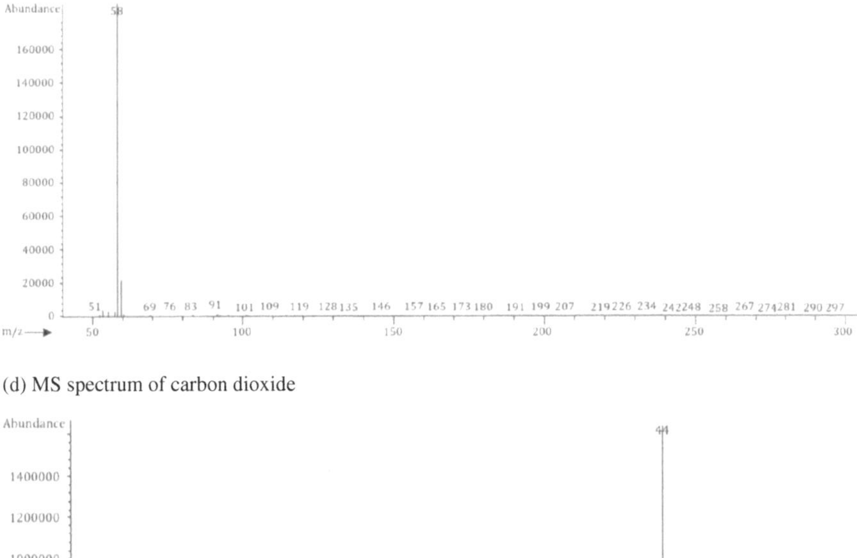

(d) MS spectrum of carbon dioxide

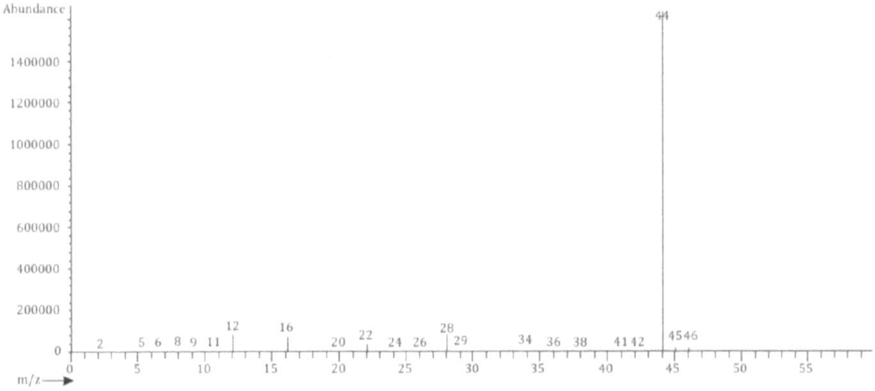

(Continued)

TABLE 2.4 *(Continued)*

(e) MS spectrum of carbon monoxide

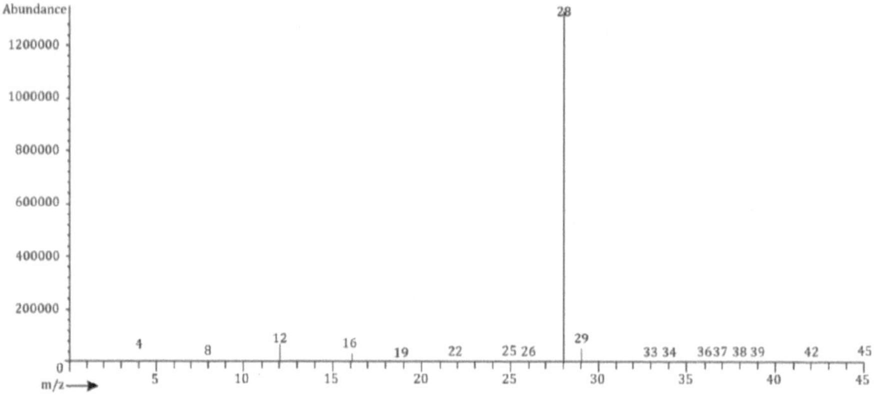

(f) MS spectrum of oxygen

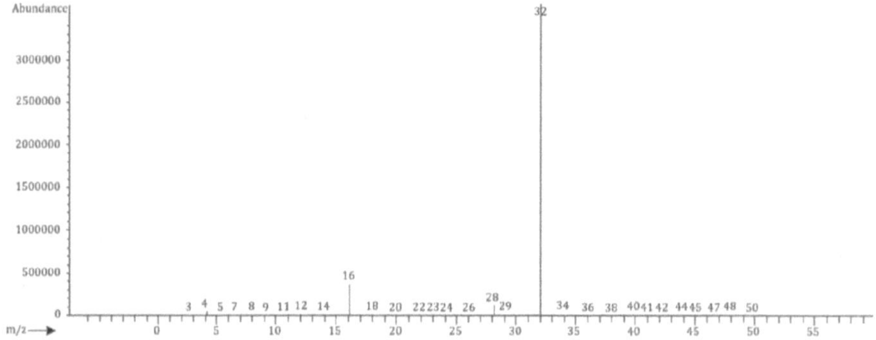

(g) MS spectrum of nitrogen

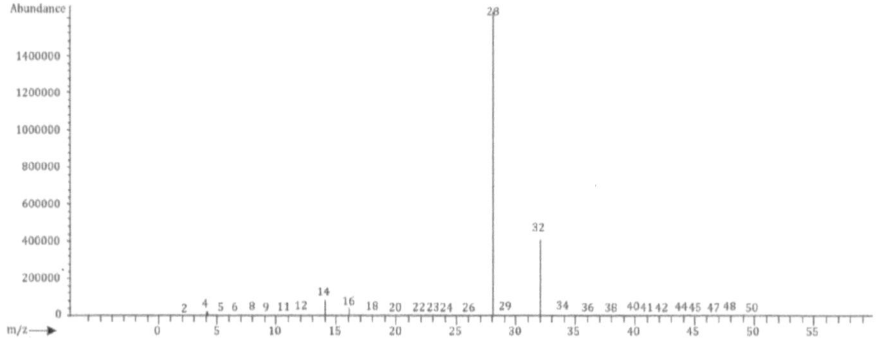

TABLE 2.4 ***(Continued)***

(h) MS spectrum for water

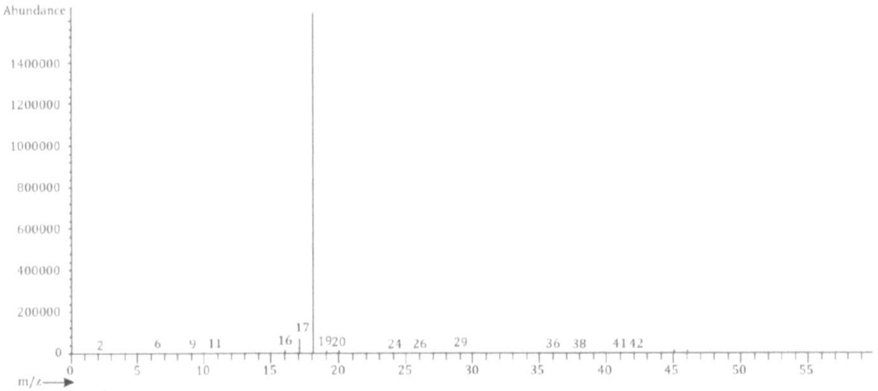

$$x_{O_2} \cong 0.209 - 1.264 \cdot x_{CO_2} \tag{2.2}$$

where x_{CO_2} is contents of carbon dioxide and x_{O_2} is contents of oxide.

In experimental conditions, water usually occurs in a vapor form,[6] but water vapor can condense into liquid water in a sampling line because it is enough long for condensation to occur (approx. 2). Therefore, some water can be removed. In the case of water vapor in reaction products, derivation of the relationship between the oxygen contents and carbon dioxide is as follows:

$$\left(CH_3\right)_2 CO + 4 \cdot O_2 \xrightarrow{catalyst} 3 \cdot CO_2 + 3 \cdot H_2O \qquad \frac{V_{O_2}}{V_{CO_2}} = \frac{V_{O_2}}{V_{H_2O}} = \frac{4}{3}$$

$$\left.\begin{array}{r} V_{O_2} \cong 0.209 \cdot V \\ V_{CO_2} \cong 0.000 \cdot V \\ V_{N_2} \cong 0.780 \cdot V \\ V_{H_2O} \cong 0.000 \cdot V \\ V_{residual} \cong 0.011 \cdot V \end{array}\right\} \xrightarrow[reaction]{cathalitic} \left\{\begin{array}{l} V_{O_2} \cong 0.209 \cdot V - V_{CO_2 + H_2O} \\ V_{CO_2} \cong \frac{3}{4} \cdot V_{CO_2} \\ V_{N_2} \cong 0.780 \cdot V \\ V_{H_2O} \cong \frac{3}{4} \cdot V_{CO_2} \\ V_{rest} \cong 0.011 \cdot V \end{array}\right.$$

$$Sum: \qquad V \qquad\qquad V - \frac{1}{3} \cdot y$$

where V is volume of reagents, V_{CO_2} is volume of carbon dioxide, and V_{O_2} is volume of oxide;

$$\begin{cases} x_{CO_2} \cong \dfrac{\frac{3}{4}\cdot V_{CO_2}}{V-\frac{1}{2}\cdot V_{CO_2}} \\ x_{CO_2} \cong \dfrac{0.209\cdot V - V_{CO_2}}{V-\frac{1}{2}\cdot V_{CO_2}} \end{cases} \Rightarrow \quad V \cong \left(\frac{0.75}{x_{CO_2}} - 0.5\right)\cdot V_{CO_2}$$

$$x_{O_2} \cong 0.209 - 1.473\cdot x_{CO_2} \tag{2.3}$$

where x_{CO_2} is contents of carbon dioxide and x_{O_2} is contents of oxide,

Figure 2.3 shows the relationship between oxygen O_2 and carbon dioxide CO_2 contents obtained from the experiments in Table 2.3.

It is evident from Figure 2.3 that there is a good correlation[7] between oxygen O_2 and carbon dioxide CO_2 contents in the post-reaction mixture. A good correlation between the experimental results and theoretical model (2.2) was recorded, as well. As it follows from the above, the post-reaction mixture contained water vapor. However, it is evident from Figure 2.3 that for 0,95 confidence level[8] there are some points lying beyond confidence level. This can suggest that in some cases, water can be at least partly condensed.

Sometimes significant carbon monoxide CO contents were recorded during the experiments (Table 2.3). The increase in the carbon monoxide CO contents was a result of the increase in the catalyst temperature in the reaction zone (Table 2.5).

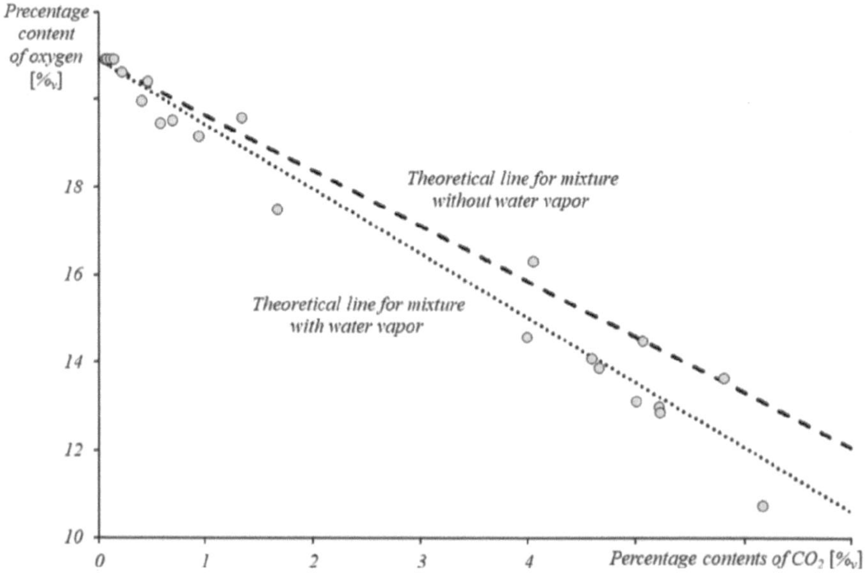

FIGURE 2.3 Comparison of measurement results from Table 2.3 with mathematical models of oxidation reaction (2.2) and reaction (2.3).

TABLE 2.5
Carbon monoxide amount in the post-reaction mixture as a function of catalyst temperature in the oxidizing zone

The mean catalyst temperature in the reaction zone			Maximum carbon oxide concentration
Mean	middle part	upper part	
	[°*C*] ±2°*C*		[*ppm*] ±2 *ppm*
474	515	433	>2000
359	360	357	350
400	400	400	1000
435	470	400	1500
355	360	350	500

This may have been caused by incomplete combustion or carbon dioxide CO_2 dissociation to carbon monoxide CO and oxygen O_2 that can occur at a high temperature on catalyzer. A rapid increase in temperature[9] in the reaction zone during the work of the reactor was observed. This made cooling of the reactor necessary. Otherwise, the oxygen uptake simulation would have continued for only 1 *min* before thermal destruction. This amount of time is insufficient to carry out research on ventilation of the SCR breathing space.

DIMENSIONALITY REDUCTION OF THE PROBLEMATIC SITUATION

The *exploratory factor analysis*[10] (EFA) technique was used to analyze the data obtained from the experiment. This method can be used to reduce the problem dimensions, which makes it is easier to determine the mechanisms[11] of chemical reaction occurring in the catalyzer.[12] In the first phase of EFA *principal component analysis*[13] PCA was used as a method for factor extraction.

As a rule, the statistical EFA is used for two purposes: to reduce the number of the problem dimensions being analyzed and to determine the problem structure. Determination of the problem structure involves transforming the existing variables into the orthogonal coordinate system and hierarchic classification. Reduction of the number of variables is applied to many technical problems when a complex measurement is needed to describe many features of the object being investigated. Reduction of the number of variables describing the problem is performed due to analytical limitations. In many cases, reduction is possible without significant loss of information. It takes place, e.g., in the case of a two-dimension problem, when data scatter diagram is similar to that presented in Figure 2.4. Of course, the example of a two-dimensional problem can be extended to multidimensional space.

It is evident from Figure 2.4 that the presented data is strongly correlated and replacement of $X \perp Y$ coordinate system with concentration ellipse axis $X' \perp Y'$ makes it possible to characterize measuring points with the use of variable X' and partial loss of information only. In this case, the reduction of dimensions will be connected with rotation of the coordinate system by 45°. At the beginning, covariance

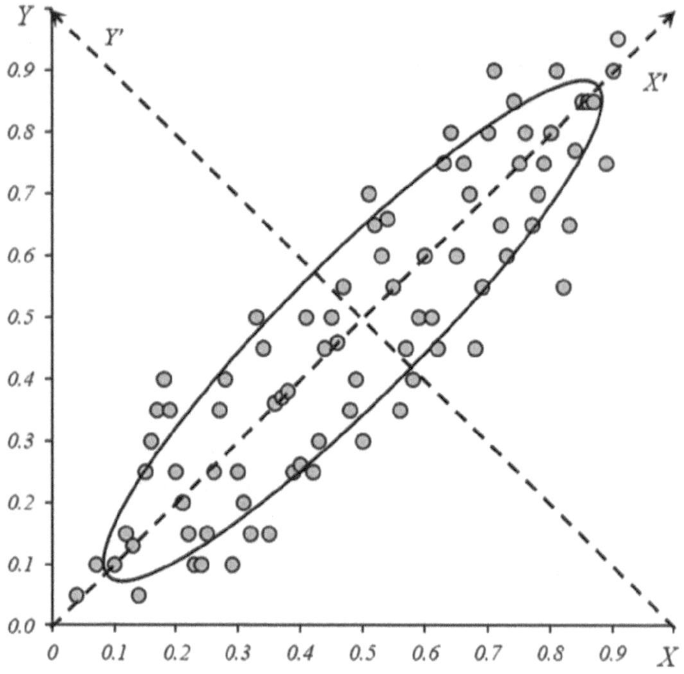

FIGURE 2.4 An example of a two-dimensional scatter plot with a selected ellipse of concentration and its main axes for translated data on standard variability in the range [0;1].

matrix that determines the mutual relationship between variables should be built. Prior to that, data standardization should be done, as data is usually expressed in the different units. It involves subtracting from each measured value its mean value and dividing the difference by the standard deviation. For the data prepared this way the covariance matrix becomes the correlation matrix. For the two-dimensional case, it will be as follows[14]:

$$\forall_{\bar{x}\equiv 0;\,\sigma\equiv 1}\ \underline{C}=\begin{pmatrix} V_x & C_{xy} \\ C_{xy} & V_y \end{pmatrix}=\begin{pmatrix} 1 & r \\ r & 1 \end{pmatrix}=\underline{r} \tag{2.4}$$

where σ is standard deviation, V is standard deviation $V \equiv \sigma^2$, C is covariance, $\bar{x}$ is mean value, $\underline{C}$ is covariance matrix, and $\underline{r}$ is correlation matrix.

The values lying beyond the diagonal of the covariance matrix (2.4) determine the mutual correlation between the values. The picture of the diagonal matrix of the covariance[15] of the coordinate system rotation by 45° was applied in this case. The rotation establishes the axes of the coordinate system along the principal axis of the scatter ellipse:

$$\underline{r}=\begin{pmatrix} 1 & r \\ r & 1 \end{pmatrix}\xrightarrow{45^\circ rotation}\begin{pmatrix} a^2 & 0 \\ 0 & b^2 \end{pmatrix}$$

It follows from the assumption that after completing the operation, the total scatter of data[16] cannot change, in this case: $a^2 + b^2 = 1^2 + 1^2 = 2$. The new variables formed from the initial variables after the rotation of the coordinate system are called the *principal components*. Due to their attributes, these variables are not correlated,[17] their mean values are equal to zero $\bar{x} \equiv 0$, and they differ from each other only by the scatter.[18] The relationship between the principal components and the original attributes can be written for the particular attribute as follows:

$$X_i = \sum_k a_{ik} \cdot X_k' \tag{2.5}$$

where X_i is the previous variable, X_i' is the new variable, i is the number of original variables, k is the number of main components, and a_{i1} is the factor loads.

An analysis that takes into account all the principal components, however, does not lead to reduction of the problem dimensions. If some principal components are neglected – for example, if equation (2.5) is valid for $k > 2$ – it can be simplified for only two new variables, as follows: $X_i = a_{i1} \cdot X_1' + a_{i2} \cdot X_2' + E_i$, where E_i represents peculiar contribution for i-th attribute.

It is convenient to subject the rest of variables X_1' and X_2' to standardization. After the standardization is executed, the variables are known as factors. Variance of the variable X_i is given as:

$$V(X_i) = a^2 + b^2 = 1^2 + V(E_i) \tag{2.6}$$

where $a^2 + b^2$ is the common variation reserves and $V(E_i)$ is the peculiar variance.

However, after the operations have been completed, the interpretation of the relationships between the attributes measured can still be difficult. To facilitate the interpretation, some strategies of simplification of the factor structure were developed. The factor loads can be treated as correlation coefficients between the factors and variables, therefore they can be repeatedly subjected to rotation performed in such a way that finally some of the values are high and the others are close to zero. Such a structure is known to be the simple structure, and it is convenient to employ this method[19] using computer software. *Varimax* rotation is the most frequently used strategy. It involves maximizing the variance for the particular attribute. After this rotation is completed, the original variables are displaced near the axis of the next factor in the space.

The standardized experimental data from Table 2.3 is presented in Table 2.6.

The correlation coefficients obtained for these data are shown in Table 2.7.

It is evident from Table 2.7 that the oxygen content is highly correlated with the carbon dioxide contents. It is apparent, as carbon dioxide is the main product and oxygen is the main substrate of the high-yield reaction. In the case of carbon monoxide and methane contents, a similarly strong correlation occurs. As a matter of fact, for the sensor fitted in the gas analyzer *Multiwarn* this is a measurement of the same substance contents. The temperature of the reactor's bottom part is strongly correlated with the temperature in its middle part, and this middle part with the

TABLE 2.6
Standardized data from Table 2.3 prepared for the factor analysis

№	Temperatures						Concentration of components in post-reaction mixture							
	bottom part		middle part		upper part		Oxygen O_2		carbon dioxide CO_2		carbon monoxide CO		methane CH_4	
	measured	standardized	measured	standardized	measured	standardized	measured	standardized	measured	standardized	measured	standardized	measured	standardized
	[°C]	[–]	[°C]	[–]	[°C]	[–]	[$\%_v$]	[–]	[$\%_v$]	[–]	[*ppm*]	[–]	[%*LEL*]	[–]
1	350	2.360	515	2.272	433	1.681	19.44	0.910	0.58	–1.021	376.98	–0.129	1.43	–0.600
2	247	–0.279	360	–0.926	357	–1.104	19.95	1.090	0.40	–1.106	40.37	–1.099	0.22	–0.862
3	247	–0.279	360	–0.926	357	–1.104	12.85	–1.408	5.22	1.169	891.43	1.355	12.06	1.698
4	237	–0.535	400	–0.101	400	0.472	19.16	0.812	0.94	–0.851	66.25	–1.024	0.27	–0.851
5	237	–0.535	400	–0.101	400	0.472	14.48	–0.835	5.06	1.093	1000.55	1.669	11.65	1.610
6	268	0.259	422	0.353	400	0.472	19.52	0.939	0.70	–0.965	65.50	–1.027	0.39	–0.825
7	268	0.259	422	0.353	400	0.472	16.30	–0.194	4.05	0.617	431.53	0.029	3.91	–0.064
8	209	–1.252	360	–0.926	350	–1.360	13.12	–1.313	5.00	1.065	500.00	0.226	3.72	–0.105
mean	257.875		404.875		387.125		16.852		2.7438		421.576		4.2062	
standard deviation	39.033		48.481		27.297		2.842		2.118		346.875		4.624	

TABLE 2.7
Correlation coefficients for standardized data presented in Table 2.6

		Contents			Temperatures		
		CO_2	CO	CH_4	t_{bottom}	t_{middle}	t_{upper}
Contents	O_2	−0.98	−0.84	−0.82	0.51	0.48	0.50
	CO_2		0.86	0.84	−0.50	−0.44	−0.41
	CO			0.96	−0.16	−0.14	−0.12
	CH_4				−0.27	−0.30	−0.23
Temperatures	t_{bottom}					0.92	0.75
	t_{middle}						0.92

temperature of the reactor's upper part, whereas the temperatures of the reactor's bottom and upper parts are not strongly correlated any longer. It is probably due to the fact that the temperature of the reactor's bottom depends only on the temperature of the gas supplied to the reactor.

In the initial reaction zone[20] an inlet gas temperature effect is also observable, but it becomes weaker in the reactor's upper part, as the reaction is highly exothermic and produces a strong effect. High values of the correlation coefficients given in Table 2.7 show that many quantities are strongly correlated with each other. It means that information supplied should be in excess, and simplification can be attempted by means of coordinate hypersystem rotation. This makes it possible to neglect some attributes of less importance. In order to do this, the correlation matrix should be diagonal[21] in such a manner that the square sum of the diagonal elements[22] should be the same as the previous one. The eigenvalues for the diagonal correlation matrix by this method are presented in Table 2.8.

Only five eigenvalues were calculated; as a matter of fact, CO and CH_4 contents are the same due to the measuring technique applied.

It is evident from Table 2.8 that the two first eigenvalues explain approximately 93% of the phenomenon. When reducing the number of variables,[23] it is necessary to leave only the components of the eigenvalues greater than 1[24] (Dąbrowski, 2000;

TABLE 2.8
Eigenvalues of diagonal correlation matrix after constrained to five main component values

№	Eigenvalue	Percentage	Cumulated percentage
1	4.466	63.80	63.80
2	2.020	28.85	92.66
3	0.256	3.66	96.31
4	0.222	3.17	99.46
5	0.023	0.33	99.31

Dobosz, 2001). When the residual correlation[25] is calculated, it is evident that the values are very small, lower than 0.01. It proves that the use of two components is satisfactory to describe the problem.

To improve the factor structure, *varimax standard* type rotation constrained to two eigenvalues of diagonal correlation matrix was done. The factor loads were calculated as well. The results are given in Table 2.9.

The first of the factor loads taken from Table 2.9 is called *advancement of reaction* due to stronger relationship between each of gas contents in the post-reaction mixture. The second of the factor loads is called *thermal effect of reaction* due to stronger relationship between temperatures measured along the reaction front. It would be evident from Table 2.9 that a CO content is related to the reaction kinetics[26] rather than to the reaction temperature.

This was confirmed by the fact that carbon monoxide occurrence was observed after a certain time in the course of the reaction. In this case, cooling of the reactor will not lead to eliminating CO from the post-reaction mixture, provided that the temperature has an impact on the catalyst effectiveness. As previously mentioned, cooling of the reactor is necessary due to a high rate of temperature increase that makes it difficult to reach the stable equilibrium state in a tested UBA. Carbon oxide occurrence during the reaction may be related to the catalyst locking or too low oxygen contents in the mixture in the reaction zone. It seems that the latter is less probable, as the correlation between oxygen and carbon oxide contents $r > |-0.839|$ is not high enough. However, it is the correlation of the experimental results obtained for the outlet gas mixture,[27] whereas some deficit of oxygen may occur in the reaction core. The large carbon monoxide concentration related to the catalyst top temperature $r > 0.994$ compared to the FA results[28] is worth considering. To solve this problem, a metal reactor with partial water cooling (Figure 2.5).

TABLE 2.9
Factor loads after *varimax standard* rotation for the two first eigenvalues of correlation matrix from Table 2.8

Variable		Factor loads	
		Advancement of reaction	Thermal effect of reaction
Concentration	O_2	0.886	0.389
	CO_2	−0.909	−0.333
	CO	−0.989	0.026
	CH_4	−0.949	−0.108
Temperatures	t_{bottom}	0.189	0.915
	t_{middle}ek	0.159	0.973
	t_{upper}	0.137	0.926
Initial variance		3.569	2.917
Contents		0.5098	0.4168

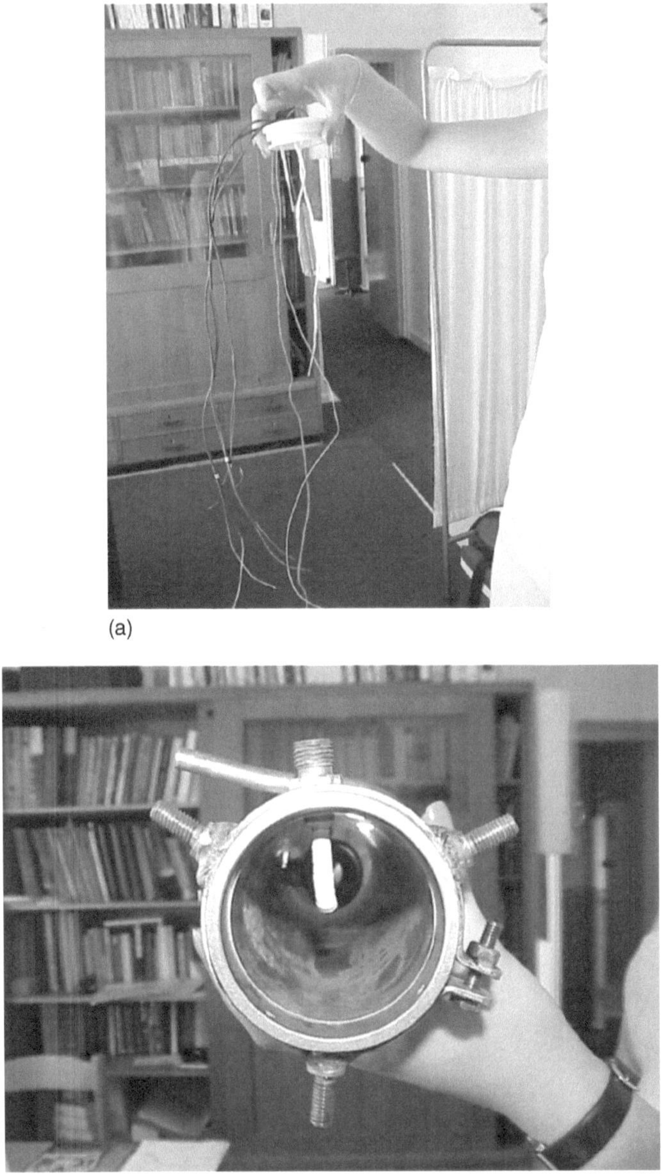

(a)

(b)

FIGURE 2.5 Metal reactor used for testing *CO* emission as the temperature function. a) Measuring sensors and heater; b) side view from the side internal of metering nozzle

(Continued)

(c)

(d)

FIGURE 2.5 (Continued) c) side view from the side internal of the water cooler; d) main view of reactor at the measuring stand.

IMPROVEMENTS AND VALIDATION

The experimental results of *CO* emission as a temperature function are presented in Table 2.10.

The results presented in Table 2.10 show that the temperature exerts a strong effect on the reaction kinetics and consequently on *CO* emission. It follows from the above that the statistical relationships previously analyzed exist in an apparent inconsistency only. As a matter of fact, the temperature exerts an effect on the catalyst

TABLE 2.10
Experimental results of carbon monoxide emission as the function of the temperature near the reaction core

t	*CO* contents [*ppm*]			
	№ of measurement			
[°*C*]	1	2	3	4
100	0	3	1	0
170				0
180	2	6	20	1
200	3	10	22	1
210	4	12	26	2
220	6	14	27	4
230	9	16	30	6
240	12	19	32	8
260	22	22	34	10
270	29	28	40	14
280				15

effectiveness. It seems that intensive internal parallel ongoing cooling of the whole reactor would be successful. On the border of the reaction core the reaction temperature should be maintained at a level of approximately 200°*C*. As it follows from the additional experiments, at temperatures less than 80°*C*, the course of reaction is weak, and below 50°*C* the reaction decays. Additionally, maintaining the temperature below 100°*C* causes the water in the reactor outlet to condense, which creates some difficulties – (Table 2.11).

Maintaining the reaction temperature at a level higher than 100°*C* requires a change of the cooling agent to ethylene glycol[29] and consequent changes in the internal cooler construction. To initiate the reaction and thermal stabilization inside, the reactor casing should be initially heated. The construction of the new reactor is presented in Figure 2.6.

TABLE 2.11
Theoretical condensation water stream emitted during acetone combustion reaction

$$(CH_3)_2CO + 4 \cdot O_2 \xrightarrow{catalyst} 3 \cdot CO_2 + 3 \cdot H_2O \qquad \varepsilon_{RQ} = 0.75$$

$3 \cdot A(C) \cong 3 \cdot 12 =$	36	$6 \cdot A(H) \cong 6 \cdot 1 =$	6
$6 \cdot A(H) \cong 6 \cdot 1 =$	6	$3 \cdot A(O) \cong 3 \cdot \times 16 =$	48
$1 \cdot A(O) \cong 1 \cdot \times 16 =$	16		
if:	58 *g* of acetone	gives	54 *g* of water
that:	73.23 cm^3 of acetone	gives	54 cm^3 of water
and:	1 cm^3 of acetone	gives	0.74 cm^3 of water

densities d: $d(acetone; 20°C) \cong 0.791\ g \cdot cm^{-3}$; $d(water) \cong 1.000\ g \cdot cm^{-3}$

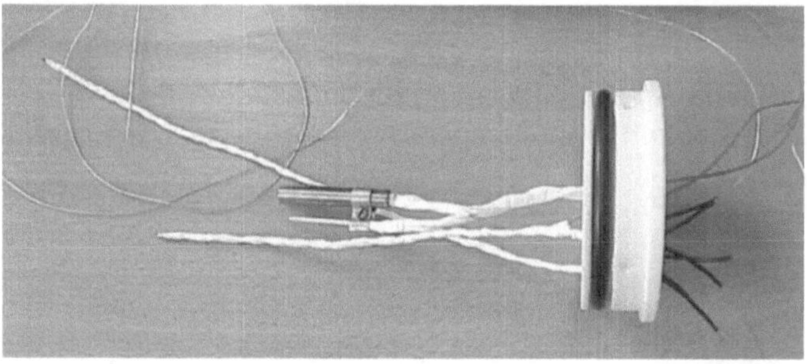

(a)

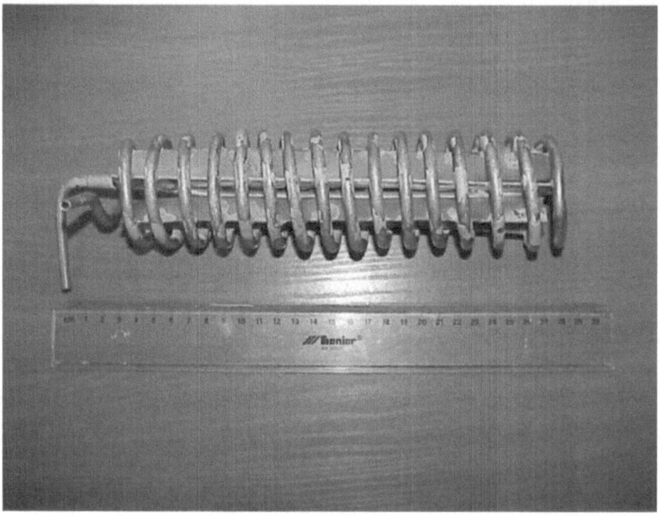

(b)

FIGURE 2.6 Laboratory stand for catalyst efficiency experiments: a) measuring equipment and internal heater; b) view of internal cooler before assembling.

The reactor shown in Figure 2.5 is filled with the catalyst to the height of approximately 15 *cm*.[30] The cooler was immersed in the catalyst layer above the dosing pipe. To test the reactor temperature, five different test runs were performed. In the test results, 290 measuring points were obtained. The temperatures were recorded by the computer system, and CO_2, CO, O_2 contents were recorded in the memory of the *Multiwarn* gas analyzer. An example of a measuring run is presented in Figure 2.7.

The temperature ranges of all the measurements of the mean gas contents obtained in the individual reactor runs, and cumulative values for all the results obtained, are presented in Table 2.12.

The mean CO_2 contents differed from the individual measurements more significantly than it followed from the standard error calculated for its cumulated population. It could have resulted from a lack of gas flow control and the fact that the different gas outlet temperatures were not taken into consideration. As it follows

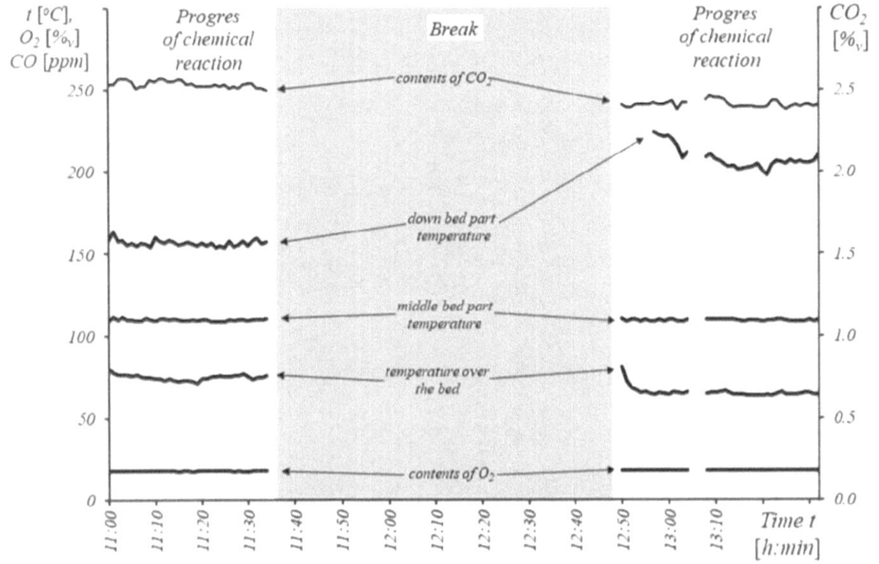

FIGURE 2.7 Example of a catalyst measuring course – measurement №5 in Table 2.12.

from the preliminary analysis of the mean values obtained in the individual measurements, the real repeatability of the measuring stand, in these conditions, is at a level of $\pm 0{,}2\%_v CO_2$. This value is placed on the border of the real measuring accuracy offered by the gas analysis methods applied.

To obtain the equations describing the relationship between O_2 and CO_2 contents, three experimental runs were carried out for extended content ranges of O_2 and CO_2. The results are presented in Table 2.13.

A good enough agreement of the experimental results obtained with the theoretical models can be noticed comparing the results with Equations. (2.2) and (2.3). Negligible differences in the slope are probably due to different moisture contents in the gas analyzed. Air was used as a breathing medium. The SCR APW-6M SCUBA constant gas metering system maintained the gas flow at a level of $(8.61 \pm 0.06) dm^3 \cdot min^{-1}$. The gas constant metering measurements were performed on the measuring stand. The experimental results[31] are presented in Table 2.14.

Apart from the oxygen contents in the combustible compounds equivalent to methane content CH_4, contents of CO and CO_2 were measured. The contents of CH_4 and CO_2 were below the measuring threshold, whereas the CO contents reached 100 *ppm*.

The tested attachment that enables oxygen uptake and CO_2 emission was connected to the breathing machine presented in Figure 2.8. In the experiments, SCR APW-6M SCUBA was used.

When the oxygen contents decrease below[32] approx. $12\%_v$, there occur problems with reaction stabilization, as the combustion process in the catalyst gets weaker and the reaction stalls. This is a typical phenomenon known from the fire thermodynamics. To elaborate on the experimental results, the same method as the method used for the preliminary model verification was applied. Such an approach was necessary, as

TABLE 2.12
Cumulative experimental results obtained from six reactor runs

		mean	−95% confidence level	+95% confidence level	median	minimum	maximum	discrepancy	variance	standard deviation	standard error
	t_{upper} [°C]				75.4	57.1	128.8	71.7			
	t_{middle} [°C]				109.9	86.2	172.4	86.2			
	t_{bottom} [°C]				166.0	141.0	224.3	83.3			
	CO [ppm]				106.9	0.0	277.1	277.1			
1	CO_2 [%$_v$]	2.2	2.2	2.2	2.2	2.1	2.3	0.2	0.00	0.033	0.004
	O_2 [%$_v$]	17.5	17.5	17.5	17.5	17.5	17.7	0.2	0.00	0.038	0.004
2	CO_2 [%$_v$]	2.1	2.1	2.1	2.1	2.1	2.2	0.2	0.001	0.037	0.005
	O_2 [%$_v$]	17.7	17.7	17.7	17.7	17.6	17.9	0.3	0.001	0.037	0.005
3	CO_2 [%$_v$]	2.2	2.2	2.2	2.2	2.1	2.4	0.2	0.005	0.072	0.014
	O_2 [%$_v$]	17.9	17.9	17.9	17.9	17.7	18.0	0.3	0.008	0.090	0.017
4	CO_2 [%$_v$]	2.5	2.5	2.5	2.5	2.3	2.6	0.3	0.005	0.070	0.009
	O_2 [%$_v$]	17.5	17.5	17.6	17.5	17.3	17.9	0.6	0.017	0.129	0.016
5	CO_2 [%$_v$]	2.5	2.5	2.5	2.5	2.4	2.6	0.2	0.004	0.066	0.008
	O_2 [%$_v$]	17.5	17.4	17.5	17.5	17.3	17.7	0.4	0.013	0.113	0.013
average	CO_2 [%$_v$]	2.3	2.3	2.3	2.3	2.1	2.6	0.6	0.025	0.159	0.01
	O_2 [%$_v$]	17.6	17.6	17.6	17.6	17.3	18.0	0.7	0.025	0.157	0.01

TABLE 2.13
The relationship between oxygen and carbon dioxide concentration in post-reaction mixture obtained for selected measurements

№	Equation	Correlation coefficient
1	$x_{O_2} \cong 0.209 - 1.434 \cdot x_{CO_2}$	0.9307
2	$x_{O_2} \cong 0.208 - 1.436 \cdot x_{CO_2}$	0.9983
3	$x_{O_2} \cong 0.208 - 1.445 \cdot x_{CO_2}$	0.9963

the metering nozzle placed in the reaction zone could change its characteristics, therefore acetone metering could be different due to the conditions established. The verification of the model was performed using Equation (2.1). Assuming that $x_s = 0{,}117\ mol \cdot mol^{-1}$, the oxygen consumption is as follows:

$$\dot{v} = \frac{0.209 - 0.117}{1 - 0.117} \cdot 8.61 = (0.90 \pm 0.07)\, dm^3 \cdot min^{-1}$$

The systematic error of oxygen consumption is determined with the total differential method:

$$\Delta\dot{v} = \left|\frac{\partial \dot{v}}{\partial x_w}\right| \cdot \Delta x_w + \left|\frac{\partial \dot{v}}{\partial x(\infty)}\right| \cdot \Delta x(\infty) + \left|\frac{\partial \dot{v}}{\partial \dot{V}}\right| \cdot \Delta \dot{V} \cong 0.07\, dm^3 \cdot min^{-1}$$

$$\left|\frac{\partial \dot{v}}{\partial x_w}\right| \cdot \Delta x_w = \frac{\dot{V}}{1 - x(\infty)} \cdot \Delta x_w \cong \frac{8.61}{1 - 0.117} \cdot 0{,}001 \cong 0.01\, dm^3 \cdot min^{-1}$$

$$\left|\frac{\partial \dot{v}}{\partial x(\infty)}\right| \cdot \Delta x(\infty) = \frac{1 - x_w}{1 - x(\infty)} \cdot \dot{V} \cdot \Delta x_w \cong \frac{1 - 0.209}{1 - 0.117} \cdot 8.61 \cdot 0.007 \cong 0.054\, dm^3 \cdot min^{-1}$$

$$\left|\frac{\partial \dot{v}}{\partial \dot{V}}\right| \cdot \Delta \dot{V} = \frac{x_w - x(\infty)}{1 - x(\infty)} \cdot \Delta \dot{V} \cong \frac{0.209 - 0.122}{1 - 0.117} \cdot 0.06 \cong 0.006\, dm^3 \cdot min^{-1}$$

Equation 9° inb Table 1.6, SCR APW-6M SCUBA data on the volume of the breathing loop from Table 1.11, and experimental relationship between the mean oxygen contents vs time make it possible to compare the mathematical model with the experimental results (Figure 2.9).

The comparison between the experimental data and SCR APW-6M SCUBA ventilation model (1.2) is presented Table 2.15.

It is worthy of note that despite the sufficient fitting of the theoretical curve and the experimental results, and the good representation of the theoretical line formed by measurement points,[33] the curve is turned relatively to the pivot point. According to the propagation of error theory, the existence of random errors is manifested by

TABLE 2.14
Experimental results of oxygen contents in the inspired gas for air supplied SCR APW-6M SCUBA

Time	Oxygen contents [%$_v$] for run №								Mean of oxygen contents
	1	2	3	4	5	6	7	8	
0	20.9	20.9	20.9	20.9	20.9	20.9	20.9	20.9	20.9
10	20.9	20.8	20.9	20.9	20.8	20.8	20.0	19.8	20.6
20	19.6	19.8	20.0	20.3	19.4	19.6	18.6	18.5	19.5
30	18.3	18.8	18.9	19.0	17.9	18.6	17.7	17.6	18.3
40	17.5	18.0	18.1	18.1	17.0	18.0	16.8	16.7	17.5
50	16.8	17.2	17.2	17.3	16.1	17.3	16.1	16.0	16.7
60	16.2	16.5	16.6	16.6	15.4	16.6	15.4	15.4	16.1
70	15.7	16.0	16.0	16.0	14.8	16.1	14.9	15.0	15.6
80	15.1	15.5	15.5	15.5	14.3	15.7	14.4	14.6	15.1
90	14.7	15.1	15.1	15.0	13.9	15.3	14.0	14.4	14.7
100	14.4	14.8	14.7	14.6	13.5	15.0	13.7	13.9	14.3
110	14.0	14.4	14.3	14.3	13.3	14.8	13.4	13.7	14.0
120	13.8	14.2	14.1	14.0	12.9	14.6	13.1	13.5	13.8
130	13.3	13.9	13.8	13.8	12.7	14.4	12.9	13.3	13.5
140	13.2	13.6	13.6	13.5	12.6	14.1	12.7	12.7	13.3
150	13.1	13.3	13.5	13.4	12.4	14.0	12.5	12.6	13.1
160	12.9	13.1	13.3	13.2	12.3	13.9	12.4	12.5	13.0
170	12.7		13.2	13.1	12.2	13.8	12.3	12.4	12.8
180	12.6		13.1	12.9	12.1	13.7	12.2	12.3	12.7
190	12.5		13.0	12.8	12.1	13.6	12.1	12.2	12.6
200	12.3		12.9	12.6	12.0	13.5	12.0	12.1	12.5
210	12.4		12.8	12.5	11.9	13.4	12.0	12.0	12.4
220	12.3		12.7	12.4	11.9	13.4	11.9	11.9	12.4
230	12.2		12.7	12.3	11.9	13.3	11.8	11.9	12.4
240	12.2		12.6	12.2	11.8	13.2	11.8		12.3
250	12.1		12.5		11.8	13.1	11.8		12.2
260	12.1		12.5		11.8	12.9	11.7		12.2
270	11.9		12.4		11.7	12.7			12.2

(a)

(b)

FIGURE 2.8 Experimental respiratory and metabolic simulator stand: a) the reactor supplying and control set; b) SCR APW-6M SCUBA connected to simulator *(Continued)*

statistical scatter that was not observed here. The existence of systematic statistical error connected with the linear effect is manifested by displacement of the experimental curve and the theoretical curve. However, here a different kind of displacement of the experimental curve and the theoretical curve is observed. This suggests inaccuracy of the theoretical model applied. The model that was

(c) (d)

FIGURE 2.8 (Continued) c) the reactor placed inside the hyperbaric chamber; d) the respiratory breathing simulator.

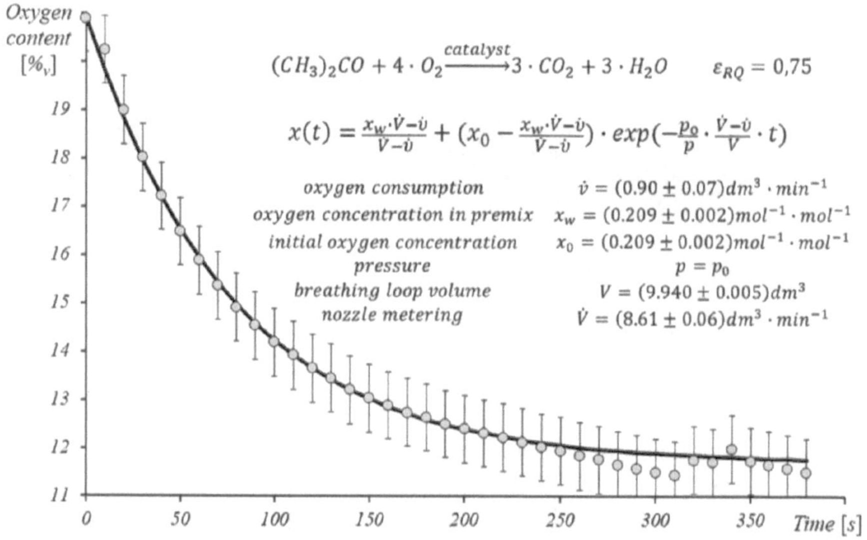

FIGURE 2.9 Comparison of oxygen contents measured in the inhalation bag of SCR APW-6M SCUBA with the mathematical model represented by a solid line.

TABLE 2.15
Comparison of oxygen contents measured in the inhalation bag of SCR APW-6M SCUBA with the mathematical model

Time	Measured oxygen content	Modeled oxygen content	Time	Measured oxygen content	Modeled oxygen content
[s]	[$\%_v$]	[$\%_v$]	[s]	[$\%_v$]	[$\%_v$]
±1 s	±0.7	–	±1 s	±0.7	–
0	20.9	20.900	140	13.3	13.653
10	20.6	19.849	150	13.1	13.481
20	19.5	18.925	160	13.0	13.330
30	18.3	18.114	170	12.8	13.197
40	17.5	17.401	180	12.7	13.080
50	16.7	16.774	190	12.6	12.978
60	16.1	16.223	200	12.5	12.888
70	15.6	15.740	210	12.4	12.809
80	15.1	15.315	220	12.4	12.739
90	14.7	14.941	230	12.4	12.678
100	14.3	14.613	240	12.3	12.624
110	14.0	14.324	250	12.2	12.577
120	13.8	14.071	260	12.2	12.536
130	13.5	13.848	270	12.2	12.499

developed earlier (Table 1.9), for the two-bag underwater breathing apparatus,[34] can be alternatively used (Table 2.15).

It seems that the conclusions drawn earlier, concerning negligible – from the practical view point – differences between models (1.2) and (1.3) in fact may not be correct, as these differences can be measured during experiments. Transformation of Eq. (1.3) with respect to oxygen consumption gives:

$$\dot{v} = \frac{\frac{p}{p_0} \cdot \frac{x_w - x_s}{x_w} \cdot \dot{V}_E + \dot{V}}{1 + \frac{p}{p_0} \cdot \frac{1 - x_s}{x_w} \cdot \frac{\dot{V}_E}{\dot{V}}} \tag{2.7}$$

where p_0 is normal pressure, p is pressure at the depth of diving, $\dot{V}$ is metering of the fresh breathing medium, $\dot{v}$ is oxygen consumption, $\dot{V}_E$ is pulmonary ventilation, x_w is oxygen molar fraction in the *premix*, and x_s is the stable value of oxygen molar fraction in the inhaled breathing bag $x_s = \lim_{t\to\infty} x(t)$.

The following experimental schedule was adopted:

- after gas mixture composition has been stabilized at the depth of 0 mH_2O, experiment was switched to the depth of 10 mH_2O
- after gas mixture composition has been stabilized at the depth of 10 mH_2O, experiment was switched to the depth of 20 mH_2O

- after gas mixture composition has been stabilized at the depth of 20 mH_2O, experiment was switched back to the depth of 10 mH_2O
- after gas mixture composition has been stabilized at the depth of 10 mH_2O, experiment was switched back to the depth of 0 mH_2O

Theoretically, calculated from Equation (2.7), the maximum difference between oxygen contents in the breathing mixture contained in the inhalation bag cannot be higher than $4\%_v$ within the depth range of 0 mH_2O–20 mH_2O. But for the same depth the oxygen contents in the breathing mixture contained in the inhalation bag (2.7) will be the same. As it follows from the above, differences in the oxygen contents should be clearly visible during an experiment planned in this manner.

The calculated oxygen consumption shown in Table 2.16 should be treated as measurements of the same parameter, and the population investigated has the normal distribution.[35]

The mean oxygen consumption is $(1.22 \pm 0.02) dm^3 \cdot min^{-1}$. The high accuracy of measurements obtained in the real experiments, including five runs for three depths, suggests a good match of the theoretical model developed for the two-bag SCR to the real object known in this case as SCR APW-6M SCUBA.

TABLE 2.16
Variation of oxygen contents in the inhalation bag and oxygen consumption resulting from it

Experiment №	Depth	Pressure increasing		Pressure decreasing	
		oxygen contents	oxygen consumption	oxygen contents	oxygen consumption
	$[mH_2O]$ $\pm 1\ mH_2O$	$[\%_v]$ $\pm 0,2\%_v$	$[dm^3 \cdot min^{-1}]$ –	$[\%_v]$ $\pm 0,2\%_v$	$[dm^3 \cdot min^{-1}]$ –
1	0	14.8	1.142	13.1	1,279
	10	12.0	1.126	–	–
	20	10.7	1.150		
2	0	13.5	1.247	12.8	1,302
	10	10.4	1.255	10.4	1,255
	20	9.4	1.255		
3	0	14.1	1.199	13.6	1,239
	10	10.5	1.247	10.7	1,231
	20	9.9	1.215		
4	0	13.4	1.255	12.9	1,295
	10	10.2	1.271	10.3	1,263
	20	9.6	1.239		
5	0	13.8	1.223	13.8	1,223
	10	11.4	1.175	11.3	1,183
	20	10.8	1.142		

SUMMARY

The two-bag SCR model (2.3) is more adequate for the presented constructions: SCR APW-6M SCUBA and SCR GAN-87 UBA. It does not rule out the earlier conclusions concerning the simplified model of one-bag SCR model (2.2). This arises from the differences in the experimental conditions. Usually, experimental conditions will be not the same as real ones because of decompression assumptions (Kłos, 2007b).

NOTES

1 *Respiratory quotient* $\varepsilon_{RQ} = \frac{\dot{v}'}{\dot{v}}$ describes the ratio of carbon dioxide output $\dot{v}'$ to oxygen intake $\dot{v}$; when carbohydrates are metabolized in the human body, it is determined at the level $\varepsilon_{RQ} = 1.0\ dm^3 \cdot dm^{-3}$, for proteins $\varepsilon_{RQ} \cong 0.8\ dm^3 \cdot dm^{-3}$, and for fats $\varepsilon_{RQ} \cong 0.7\ dm^3 \cdot dm^{-3}$.
2 0.5% platinum deposited on ca. 3 *mm* diameter aluminum rollers.
3 *Pellistor* is a solid-state thermal resistor sensor used to detect gases that are either combustible or have a significant difference in thermal conductivity to that of around atmosphere.
4 The sensor is carbon oxide sensitive, but it is scaled to a percentage of the lower explosive limit *LEL* for methane (Table 2.2).
5 Correlation coefficient $r > 0.96$.
6 Because of high reaction temperature.
7 Correlation coefficient $r > |-0{,}98|$.
8 Confidence level is not marked in the figure.
9 Approx. 3 *min*.
10 EFA is a statistical method used to describe variability among observed, correlated variables in terms of a potentially lower number of unobserved variables called factors.
11 Therefore an unfavorable *CO* production mechanism.
12 Catalyzed acetone oxidation process.
13 PCA is a statistical procedure that uses an orthogonal transformation to convert a set of observations of possibly correlated variables into a set of values of linearly uncorrelated variables called principal components.
14 For standard values $\forall_{\bar{x}=0;\ \sigma=1}\ r \equiv \frac{C_{xy}}{\sigma_x \cdot \sigma_y} \Rightarrow r = C_{xy}$.
15 Correlation.
16 Equals the sum of all variances in the data.
17 Are orthogonal.
18 Variance.
19 Strategies.
20 In the middle.
21 The matrix values exist only on the matrix diagonal.
22 The sum of variance.
23 Reduction of the problem dimension according to *Kaiser's criterion*.
24 Components №1 and №2 from Table 2.8.
25 The difference between the correlation coefficients of the initial matrix and the correlation coefficient calculated from the factor values.
26 Advancement of reaction.
27 Averaging value.
28 *CO* concentration is stronger related to the reaction kinetic.
29 Boiling point approx. 200°*C*.
30 Reactor diameter is the same as previously used –70 *mm*.
31 The breathing gas composition recorded by *Multiwarn* gas analyzer.

32 Meaning at sufficiently large acetone dosage.
33 The curve is contained within the limits of calculated maximum systematic errors.
34 As a matter of fact, the diving apparatus type APW-6M is a two-bag SCR.
35 *Liliefor's*, *Shapiro's-Wilk's,* and *K-S* tests.

REFERENCES

Dąbrowski J. 2000. *O problemie redukcji wymiarów*. Kraków: Polskie Towarzystwo Inżynierii Rolniczej. 83-905219-4-6.

Dobosz M. 2001. *Wspomagana komputerowo statystyczna analiza wyników badań*. Warszawa: Akademicka Oficyna Wydawnicza EXIT. ISBN 83-87674-75-3.

Dräger Safety AG & Co. KGaA. 2003. *Multi-Gas Monitor Multiwarn II Technical Handbook*. Lübeck: Dräger Safety AG & Co.

Frånberg O. 2015. Oxygen content in semi-closed rebreathing apparatuses for underwater use: Measurements and modeling. PhD diss., Stockholm School of Technology and Health. ISSN 1653-3836.

Kłos R. 2002. Metabolic simulator supports diving apparatus researches. *Sea Technology* 12, 53–56.

Kłos R. 2007a. Modelowanie procesów wentylancji obiektów normo- i hiperbarycznych. Higher PhD diss., The Polish Naval Academy. PL ISSN 0860-889X nr 160A.

Kłos R. 2007b. *Mathematical modelling of the normobaric and hyperbaric facilities ventilation*. Gdynia: Polish Hyperbaric Medicine and Technology Society. ISBN 978-83-924989-0-2.

Loncar M. 1992. Breathing simulator simulates human oxygen intake. *Offshore* 10, 82.

3 Manned Research of SCR with Constant *Premix* Dosage

Consistence of the assumed working parameters with the real parameters is tested with the same method as the stabilization and homogenization of the breathing gas composition for SCR APW-6M SCUBA. As a rule, diving cycles and unmanned investigations are used.

PRELIMINARY TESTING

The preliminary testing of the mathematical model of the breathing space in a diving apparatus was performed in normobaric conditions. The preliminary operating conditions for the SCR APW-6M SCUBA diving apparatus are presented in Tables 1.11–1.12 and Figure 1.7. The experiments carried out confirmed the assumptions (Table 1.14).

DISTANCE SWIMMING AT A SMALL DEPTH

Swimming in a 50 *m*-long swimming pool over distances of 50, 100, 200, 400, and 800 *m* was done at a depth not exceeding 1.1 mH_2O at maximal possible speeds adopted for a diver equipped with swim-fins. The swimming time was measured at specific distances. The divers had to try to maintain their efforts at a level that did not produce any disturbances in breathing; if this was not the case, the experiment was to be interrupted.

The oxygen contents were measured in the breathing bag of the diving apparatus directly after each of the assumed swimming distance. An oxygen analyzer, *Servomex* type, was used.[1] The part of experiments aimed at verifying these assumptions is presented in Figure 3.1.

The experiments are part of investigations aimed at confirming the design assumptions. Detailed data were published earlier in the literature (Kłos, 2007a; Kłos, 2007b). According to the medical and technical assumptions, oxygen contents in the inspired breathing mixture should not go below the level of $20\%_v$ during experiments. The experiments carried out confirmed the assumptions (Table 1.14).

PRESSURE TESTS WITHOUT IMMERSION OF THE DIVER

In order to verify the correctness of the design assumptions, pressure tests without immersion of the diver[2] were carried out. Exercise tests were performed approximately 5 *min* after pressurization of the diving complex. The temperature during the tests was maintained at a level of 20–22 ± 1°*C*. The diver breathed *Nitrox Nx* mixture[3]

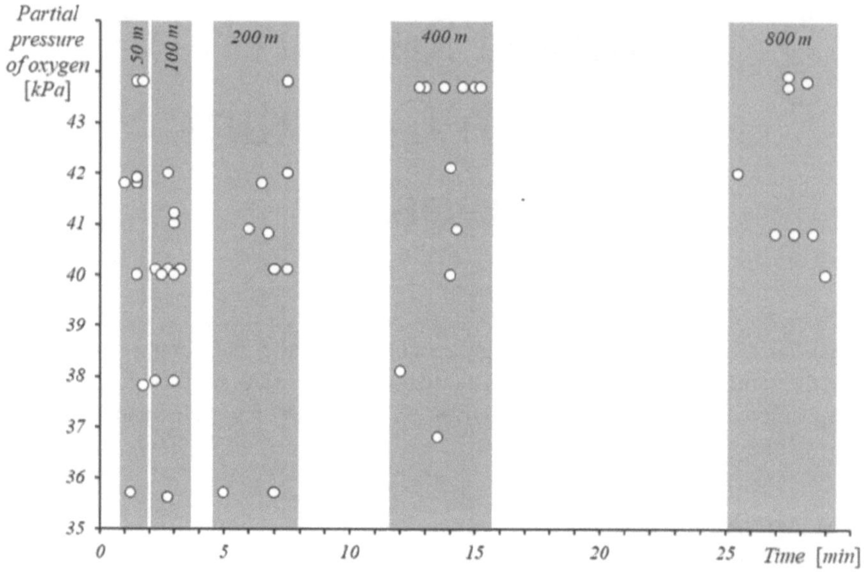

FIGURE 3.1 The experimental results of oxygen contents in the inspired breathing mixture for SCR APW-6M SCUBA during swimming at distances of 50, 100, 200, 400, and 800 *m* at low depth.

containing (28 ± 0.5)$\%_v$ of oxygen contained in the diving apparatus (Table 1.14). During the experiments the divers were equipped with a hose fitted to the type SCR GAN-87 diving apparatus. During the experiments the subjects performed exercises at the work rate *of* 2 *W* per 1 *kg* of body mass. In order to avoid inspiration of air from the chamber, the divers used nose clips. The divers were not loaded with the diving apparatus.[4] The divers exercised using a type *Monark* bicycle ergometer. The accuracy of the work rate was ±5%. The subjects exercised for 10 *min*. After each 10 *min.* cycle, the subjects rested. The 40 *min* period of diving consisted of two 10 *min.* exercise cycles. During a 1 *h* period of diving the divers performed two cycles lasting 10 *min* and 15 *min*. The experiments were performed at different depths of diving, from 30 mH_2O to 80 mH_2O; duration of exposition was maintained at a level of 40 *min* and 60 *min*. Oxygen percentage contents in the inspiration bag were measured by means of the *Servomex* analyzer with maximal absolute measurement error of ±0.5$\%_v$. Oxygen percentage was measured every 1 *min*. In order to confirm the measurement correctness, the oxygen contents were measured every 5 *min.* with a type CDS 111 *Varian Aerograph* chromatograph with maximal absolute measurement error of ±0.2$\%_v$. In order to control the functioning of carbon dioxide scrubber, a continuous check control of carbon dioxide contents in the inspiration bag was performed with a *Infralyt* 4 gas analyzer within the range 0–0.1 ± 0.01$\%_v CO_2$. The oxygen and carbon dioxide contents were measured outside the hyperbaric chamber after expansion of the gas mixture studied.

A special collector was used to sample the gas mixture from the inspiration bag. It enabled us to analyze the gas mixture by means of a gas analyzers placed outside

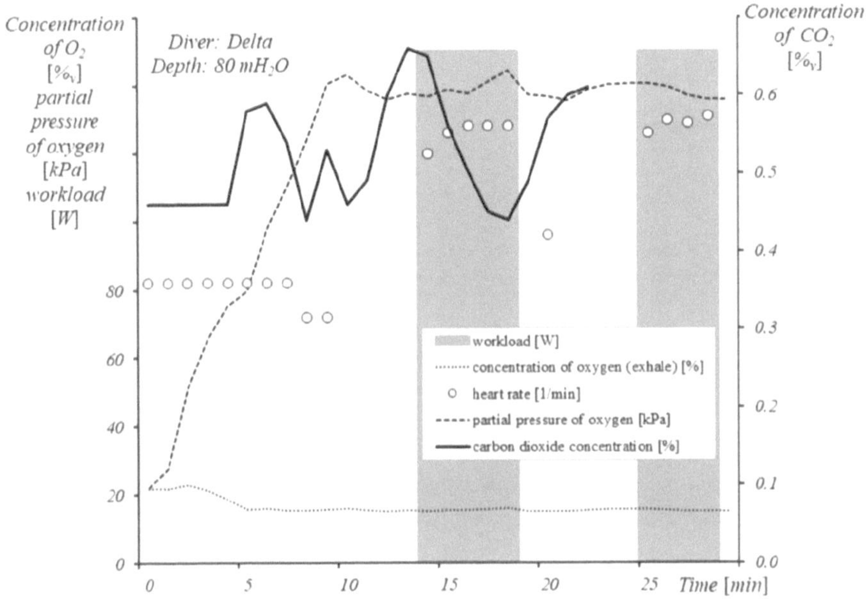

FIGURE 3.2 Example of oxygen contents measurements in the inspired gas mixture from the UBA type SCR GAN-87 during exercise on the cycle ergometer in the dry hyperbaric environment.[5]

the hyperbaric chamber. The delay time caused by the use of a measuring line was from 2 *min* to 3 *min* due to the depth of diving. The sample was the smallest possible amount of gas mixture from the breathing bag. The stream of the sampled gas was maintained at a level not exceeding 2 $cm^3 \cdot min^{-1}$. The delay of the analytical line was taken into account in the presentation of the results. A *System* 8000 monitoring set produced by *Biazet* was used to measure the heart rate with the telemetry method. The example results are presented in Figure 3.2. Detailed data was published earlier (Kłos, 2007a; Kłos, 2007b).

The results presented are part of the investigations carried out to confirm the design assumption (Table 1.14). According to the medical[6] and technical assumptions in Table 1.14, oxygen partial pressure in the inspired gas mixture during the diving tests should not exceed the range $p_i \in [20, 160]kPa$.

PRESSURE TESTS IN IMMERSION

The pressure tests carried out in the swimming pool[7] inside the land-based diving complex were the next stage of research carried out to verify the design assumptions.

The divers swam at the sub-maximal work rate. The divers tried to maintain work rate at the level that did not produce breathing disturbances; if this was not the case, the experiment was to be interrupted. The subjects started swimming 5 *min* after the pressurization of the chamber. During the experiments, the divers were equipped

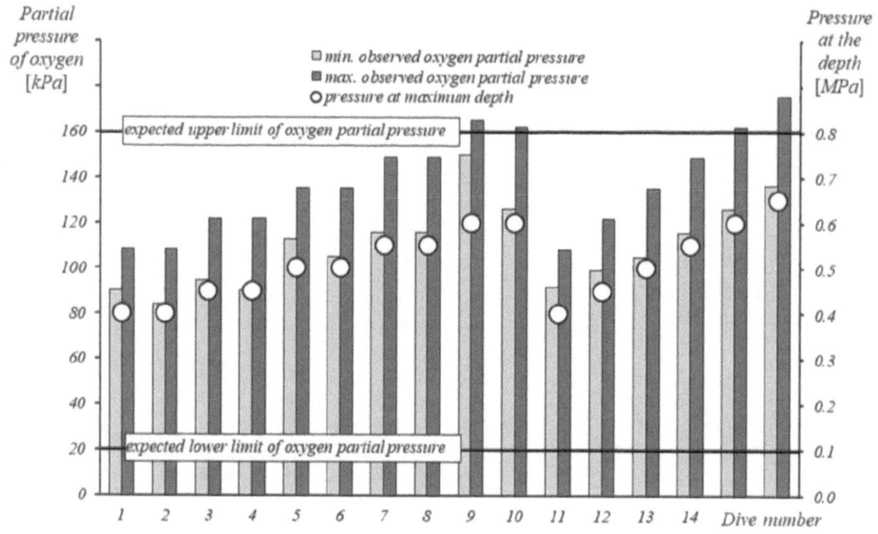

FIGURE 3.3 Setting up the minimal and maximal oxygen partial pressures obtained in the diving experiments with the use of the diving apparatus SCR GAN-87 UBA.

with the SCR GAN-87 UBA diving apparatus. The diving apparatus had a hose supplied with *Nx* gas mixture containing $[28 \pm 0.5]\%_v$ of oxygen. The experiments were carried out at different diving depths from $[30;56]mH_2O$. All measurements were taken as in the previous tests, the pressure tests without immersion. The example results are similar to those presented earlier. Detailed data were published earlier (Kłos, 2007a; Kłos, 2007b).

Figure 3.3 shows a comparison of the oxygen partial pressures in the inspired gas recorded in the diving experiments with the use of SCR GAN-87 UBA during simulated diving and diving in the wet-pot as well. It is evident from dive №15 that the permissible upper oxygen partial pressure limit was reached and slightly exceeded during diving at the depth of 50 mH_2O.

It is evident from dive №16 that the permissible upper oxygen partial pressure limit was significantly exceeded during diving at the depth of 55 mH_2O. The lower oxygen permissible partial pressure limit was not exceeded during any of the dives.

STATISTICAL PROCESSING OF THE RESULTS

The following assumption was made at the verification of the mathematical model describing ventilation of the breathing space of the semi-closed-circuit diving apparatus with constant breathing gas dosage:

- 66 tests aimed at verification of the lower oxygen partial pressure in the inspired breathing medium were carried out and all tests confirmed the assumed model;
- 16 tests aimed at verification of the upper oxygen partial pressure in the inspired breathing medium were carried out and all tests confirmed the assumed model.

For dives №8 and №9 in Figure 3.3, a slight excess of the oxygen permissible partial pressure is included within the range of the measurement error. The measured values are close to the maximal upper permissible limit. This can confirm the precision of the assumed mathematical model. Distinct excess of the permissible maximal oxygen partial pressure during dive №16 in Figure 3.3 also confirmed the assumed model. Excess of the assumed permissible depth limit of diving by 5 mH_2O also showed that the assumed mathematical model was precise. As it follows from the above, there are 82 events that confirmed the assumed mathematical model and no events negating the model.

Assuming the simple binomial distribution[8] for parent population and the samples of this population at probability $\rho = 0.05$ of the event negating this model, the probability that this kind of event will occur for the sample $N = 82$ at $n = 82$ positive events and $m = 0$ negative events equals:

$$\forall_{N=m+n;\rho+\sigma=1} P = \frac{N!}{n! \cdot m!} \cdot \rho^m \cdot \sigma^n \Rightarrow P = \frac{82!}{82! \cdot 0!} \cdot 0.95^{82} \cdot 0.05^0 \cong 0.015 \qquad (3.1)$$

where m is the quantity of negative events, n is the quantity of positive events, N is the quantity of the sample elements, P is the total conditional probability of an events combination,[9] ρ is probability of a negative event, and σ is probability of a positive event.

On the assumption that the probability of the negative event is at the level of 5%, the probability of getting the same result as in the experiments performed is at the level of 1.5%. As it follows from the above, there is low probability to obtain such a result by accident. In order to determine the range of probabilities of events confirming the correctness of the proposed mathematical model of semi-closed-circuit diving apparatus SCR with the constant dosage of the breathing medium, after the associative event described above, for which the assumption of the arbitrary event combination occurring is equal to or less than significance α, the following formula can be written[10]:

$$\begin{cases} P_{p>p_l} = \sum_{x=n}^{N} \binom{N}{x} \cdot p_l^x \cdot (1-p_l)^{N-x} = \alpha \\ P_{p<p_r} = \sum_{x=0}^{n} \binom{N}{x} \cdot p_r^x \cdot (1-p_r)^{N-x} = \alpha \end{cases} \qquad (3.2)$$

where α is significance, p_l is the left limit of the conditional probability of a combination of events, and p_r is the right limit of the conditional probability of a combination of events.

The confidence level L expressed by conditional probabilities can be written as:

$$L = \begin{cases} P(p_l < p < p_r | 0 < n < N) = 1 - 2 \cdot \alpha \\ P(p < p_r | n = 0) = 1 - \alpha \\ P(p > p_l | n = N) = 1 - \alpha \end{cases} \qquad (3.3)$$

where L is confidence level.

Hence from (3.3) for $n = N, p_r = 1$:

$$p_l = 10^{\frac{log 0.05}{N}} = 10^{\frac{log 0.05}{82}} \cong 0.96 \quad (3.4)$$

The probability of occurrence of a negating event, provided that the above associative event occurs, is at the level 0–4%. On the common assumption that the significance level in technique is 5%, the probability of negating event occurrence is lower. In other words, there is no basis for rejecting the hypothesis of ventilation model correctness of an SCR with the constant gas dosage at the confidence level of 95%. As can be seen, in order to confirm the hypothesis of the mathematical model acceptance, the minimum number of the necessary tests was carried out.[11]

The aim of the work was to develop and verify the ventilation mathematical models of an SCR with a constant dosage of the breathing medium. The models concerned are of analytical character; therefore, they have physical interpretation.

It seems that the best method for the experiments is to use a breathing machine simulating the breathing action with simultaneous oxygen sampling from the breathing space and carbon dioxide emission. Carbon dioxide emission is maintained at the proper level.[12] In such a case, there would be no necessity to carry out manned experiments, and the experimental results would be repeatable with the known accuracy of reproduction. Within the framework of the financial support of the project, a laboratory set began to be constructed. Now, there is a newer experimental set of this type, described later.

In order to verify the mathematical model of the SCR with constant breathing gas dosage, a series of experimental dives was carried out. To test the mathematical model at the upper and lower limits of oxygen partial pressures in the inspired breathing gas, two different methods were applied. The lower limits were verified during the basin tests with the use of SCR APW-6M SCUBA. Adequacy of the assumed mathematical model at upper limits of oxygen partial pressures was tested during the pressure tests with the use of the diving apparatus type SCR GAN-87 UBA. The experiments were carried out in parallel with the verification experiments of the decompression procedure. There were more similar trials carried out, but the system was modified.[13] That is why the results obtained are sufficient enough to draw conclusions concerning the adequacy of the mathematical ventilation model.

RESULTS OF EXPERIMENTS

The preliminary operation-related assumptions for the design of the SCR APW-6M SCUBA and SCR GAN-87 UBA types are given in Table 1.14. The table was based on a multiple solution to the equation set (1.10). For the SCR APW-6M SCUBA, it was assumed that the fresh breathing medium metering is maintained at a level of $8\ dm^3 \cdot min^{-1}$ for two operational gas mixtures $Nx\ 55\%_v O_2/N_2$ and $Nx\ 45\%_v O_2/N_2$. The verification of the assumptions was presented. It was carried out during distance swimming at small depths. According to the medical[14] and technical requirements of these trials, the oxygen contents in the inspired gas should not be lower than $20\%_v$. The experiments carried out confirmed these assumptions.

The pressure tests performed confirmed the above assumptions, despite the fact that the selected metering rates were lower than those obtained from the calculations. The drop in the oxygen partial pressure observed in the breathing space was compensated for by an increase in decompression time. This was clearly visible during the dives when the divers performed intensive exercises within the depth range of 20–30 mH_2O. The divers breathed the breathing mixture having the composition Nx 45$\%_vO_2/N_2$. The observed drop in the oxygen partial pressure was consistent with the proposed mathematical model of ventilation. The choice of the lower fresh breathing medium rate gave a possibility of extending the SCR protective functioning time.[15] The experimentally verified guaranteed protective functioning time of carbon dioxide scrubber is 3 h at water temperature not less than 5°C.

The preliminary operation-related data for training use of the diving apparatus type SCR GAN-87 UBA supplied by Nx are presented in Table 1.14. The working modes of the multi-nozzle SCR GAN-87 UBA are different from those featured in the SCR APW-6M SCUBA. The working mode of the apparatus is depth dependent. According to the tactical and technical assumptions, the training version of the apparatus should allow diving to average depths of up to[16] 50 mH_2O. In order to prepare the diving in compliance with the calculations presented in Table 1.14, it is necessary to use the gas mixture having the oxygen contents of approximately Nx 27$\%_vO_2/N_2$ for the range of diving depth 0–50 mH_2O or approximately Nx 28$\%_vO_2/N_2$ for the depth range 10–50 mH_2O. For these solutions the fresh breathing medium metering rates differ by 100%. For this reason, it was assumed that the metering rate was equal to the mean average of the values 40 $dm^3 \cdot min^{-1}$ and 16 $dm^3 \cdot min^{-1}$. Some of the pressure tests that were performed without immersion of the diver and those with immersion are presented above. Investigations of the diving apparatus were performed at the following parameters: *premix* Nx 28$\%_vO_2/N_2$ and gas metering 25 $dm^3 \cdot min^{-1}$. The choice of the parameters was based on the fact that the function of gas metering vs the composition of breathed mixture is weaker than the function of the fresh breathing gas composition vs the composition of breathed mixture. Maintaining oxygen partial pressure in the inspired breathing gas within the limits 20–160 kPa is a necessary condition for the experimental confirmation of the previous assumptions. For deeper dives the approach was the same, and tactical data for presented *SCRs* are presented in Tables 1.1, 1.3, and 1.4.

SUMMARY

The experiments performed show that the assumptions concerning oxygen partial pressure in the inspired breathing medium were confirmed. In this way, there is no basis for rejecting the hypothesis of the ventilation model correctness of an SCR with constant gas dosage at the confidence level of 95%.

NOTES

1 Maximal absolute measuring error ±0, 5$\%_v$.
2 Simulated diving.
3 Oxygen–nitrogen gas mixture.
4 The diving apparatus was placed on the table in front of the diver.

5 Without immersion in water.
6 Decompression.
7 Wet pot.
8 Bernoulli distribution.
9 Significance α.
10 Left and right limit of the conditional probability for any evidence sequence.
11 Experiment had to be planned economically because of the high validation costs.
12 Metabolic breathing simulator.
13 The parameters assumed at derivation of the mathematical model.
14 Decompression.
15 At the same amount of the breathing medium from an integral unit of supply flasks.
16 According to tactical and technical assumptions of the SCR GAN-87, *Nx* gas mixture should be the training *premix* and the UBA should work as an SCUBA in this case.

REFERENCES

Kłos R. 2007a. Modelowanie procesów wentylancji obiektów normo- i hiperbarycznych. Higher PhD diss., The Polish Naval Academy. PL ISSN 0860-889X nr 160A.

Kłos R. 2007b. *Mathematical modelling of the normobaric and hyperbaric facilities ventilation.* Gdynia: Polish Hyperbaric Medicine and Technology Society. ISBN 978-83-924989-0-2.

4 Ventilation of a Construction with a Metering Bellows Dispenser

In Chapter 1, older research results were presented. They are from the experimental investigations of the German SCR-UBA type FGG III semi-closed-circuit rebreather with constant dosage of *premix* and prototypes of Polish SCR-UBA type GAN-87 with constant dosage of *premix* and a self-contained version APW-6M SCUBA (Kłos, 2000).

In this chapter the results of newer scientific research investigations are presented, concentrated on the French SCUBA with alternatively closed and semi-closed-circuit SCR/CR type AMPHORA (Kłos, 2012). Recent studies are still continued and they are concerned with research of the French SCR-SCUBA type CRABE (Kłos, 2002; Kłos, 2011; Kłos, 2016).

CONSTRUCTION

The CRABE is a SCR with two breathing bags placed one inside the other. It is powered by *Nitrox Nx*, *Heliox Hx*, *Trymix Tx* as *premix,*[1] or pure oxygen[2] (Kłos, 2000).

The proper circulation of the breathing mixture is maintained by the directional valves (1) (Figure 4.1).

During the expiration phase, the exhaust valve of the mouthpiece opens (Figure 4.1A). The breathing mixture exhaled through the mouthpiece (12), the exhalation hose, and the exhaust valve passes through a carbon dioxide scrubber (11) into to the large breathing bag (2), and from there through the non-return valve into the small breathing bag (2).

When the diver inhales, the exhaust valve closes, and the inlet valve opens (Figure 4.1B). The breathing mixture is inhaled into the lungs from the large bag (2), through the inlet valve, inhalation hose, and mouthpiece (12). This causes the large bag (2) to shrink, along with the small bag (3), from which the gas escapes through a relief valve (4) into ambient water. As the large bag (2) is shrinking, it triggers the dosing valve (5) by pressing its lever. Opening the valve (5) causes fresh breathing mixture to be inhaled from the cylinder (10) through the valve (9), coupler (8), regulator valve (6), and dosing valve (5) to the large bag (2), where it is mixed with regenerated breathing mixture. From here, the breathing mixture is inhaled by the diver through the non-return valve, inhalation hose and mouthpiece (12). The demand for breathing mixture is regulated by breathing and the ratio of the volume of the large bag to the small one.

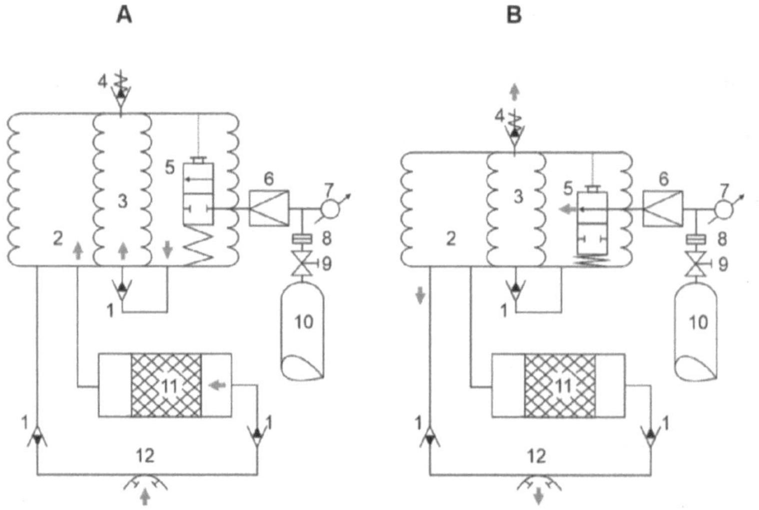

Figure 4.1 Principle of operation of a semi-closed-circuit rebreather with breathing bags placed one inside the other. A) Expiration phase, B) inspiration phase, 1 – non-return valve, 2 – external large bag, 3 – inner small bag, 4 – relief valve, 5 – dosing valve, 6 – regulator valve, 7 – manometer, 8 – coupling, 9 – shutoff valve, 10 – cylinder with breathing mixture, 11 – carbon dioxide scrubber, 12 – mouthpiece.

VENTILATION

In a SCR construction with a metering bellows dispenser, like previously, there is a relative decrease in the oxygen content to the stable value x_s in the breathing space in relation to the oxygen content in the *premix*[3] delivered to this space x_w.

The breathing gas that remains in circulation is always a breathing mixture, because even during oxygen decompression the diver does not breathe pure oxygen, as in the circuit there is always some circulating nitrogen N_2 or helium *He* left from the breathing gas, which comes from places that are not ventilated sufficiently in the breathing space of the apparatus and from the diver's tissue cleansing with oxygen. The assessment of the effectiveness of cleansing of the breathing space and the diver's body with oxygen is crucial for the correct design of safe decompression.

BALANCE

During inspiration, part of the breathing mixture is taken: $-V_I$. The elementary volume of inspiration $-dV_I = -dh \cdot (A - a)$ will be the product of the elementary change in the height of bags $-dh$ multiplied by the area of the base of the large bag A minus the area of the small bag's base area $-a$ (Figure 4.2).

At this time, the elementary drop in the height of bags $-dh$ will cause a release of the content of the small bag dV_u to water, equal to the product of the elementary drop in height $-dh$ and surface area of the small bag a: $-dV_u = -dh \cdot a$. The ratio of elementary volume of the gas released from the small sack to water $-dV_u$ to the elementary volume of inspiration $-dV_I$ will be: $\frac{-dV_u}{-dV_I} = \frac{a}{A-a} = \frac{u}{U-u} \equiv r$, where u represents

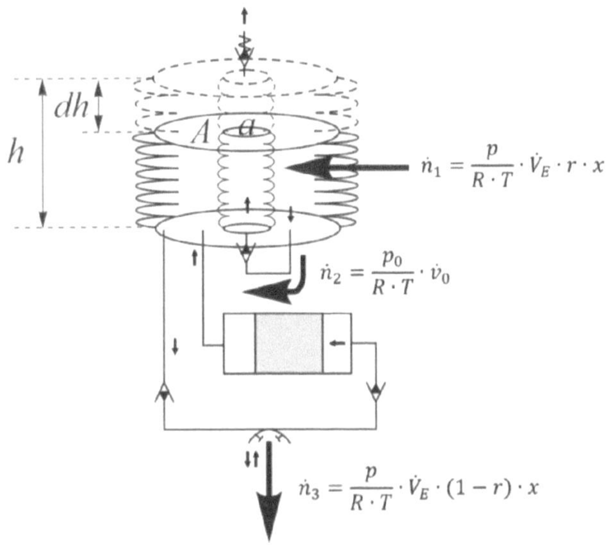

Figure 4.2 Diagram of operation of the bag system in the bag during inspiration. A – area of the base of the large bag, a – area of the base of the small bag, dh – elementary change in height of bags, h – standard height of bags, p – pressure the diving depth, p_0 – normal pressure, R – universal constant of gas, T – absolute temperature, U – volume of the large bag, u – volume of the small bag, r – volume ratio of the small bag to the large bag $r = \frac{u}{U-u}$, $\dot{V}_E$ – ventilation of lungs, $\dot{\upsilon}_0$ – stream of consumed oxygen, x – molar fraction of oxygen in the fresh breathing gas, x_w – molar fraction of oxygen in the fresh breathing gas, x_0 – molar fraction of oxygen in the breathing space before the diving apparatus is started.

the volume of the small bag and U the volume of the large bag. It follows from the above considerations that the released number of moles of the breathing mixture and, as a consequence of oxygen $-n_1 = f\left(\dot{V}_E, r\right)$, is associated with lung ventilation[4] $\dot{V}_E$ by means of the volume ratio of small bag u and large bag U: $r = \frac{u}{U-u}$. Thus, the molar stream of oxygen escaping through the small bag relief valve $-\dot{n}_1$ can be estimated using *the Clapeyron equation* as: $-\dot{n}_1 = -\frac{p}{R \cdot T} \cdot \dot{V}_E \cdot r \cdot x$, where p represents the pressure at the diving depth H: $p = f(H)$. The same number of moles $-\dot{n}_1 = -\frac{p}{R \cdot T} \cdot \dot{V}_E \cdot r \cdot x$ decreases during inspiration from the large bag to the small bag (Figure 4.2 and Table 4.1).

During inspiration, the oxygen stream having value $-\dot{n}_2 = -\frac{p_0}{R \cdot T} \cdot \dot{\upsilon}_0$ is consumed from the breathing mixture drawn in. In the same phase the dosing system resupplies the volume of the large bag with fresh *premix*, delivering oxygen having the value: $\dot{n}_3 = \left(\frac{p}{R \cdot T} \cdot \dot{V}_E \cdot r + \frac{p_0}{p} \cdot \dot{\upsilon}_0\right) \cdot x_w$. Thus, the elementary change in oxygen content ∂x in volume of the large bag $(U-u)$ at time ∂t can be written as: $\frac{p}{R \cdot T} \cdot (U-u) \cdot \frac{\partial x}{\partial t} = \dot{n}_3 - \dot{n}_2 - \dot{n}_1$

Table 4.1
Balance of streams of mixture and oxygen mass[5] for a semi-closed rebreather with a system of bags placed one above another

Particulars	Mixture Escapes	Adds	Oxygen Escapes	Adds
Inspiration	$\frac{p}{R\cdot T}\cdot\dot{V}_E\cdot(1-r)$	—	$\frac{p}{R\cdot T}\cdot\dot{V}_E\cdot(1-r)\cdot x$	—
To small bag	$\frac{p}{R\cdot T}\cdot\dot{V}_E\cdot r$	—	$\frac{p}{R\cdot T}\cdot\dot{V}_E\cdot r\cdot x$	—
Addition valve	—	$\frac{p}{R\cdot T}\cdot\dot{V}_E\cdot r+\frac{p_0}{R\cdot T}\cdot\dot{v}_0$	—	$\frac{p}{R\cdot T}\cdot\left(\dot{V}_E\cdot r+\frac{p_0}{p}\cdot\dot{v}_0\right)\cdot x_w$
Expiration	—	$\frac{p}{R\cdot T}\cdot\dot{V}_E\cdot(1-r)-\frac{p_0}{R\cdot T}\cdot\dot{v}_0$	—	$\frac{p}{R\cdot T}\cdot\dot{V}_E\cdot(1-r)\cdot x-\frac{p_0}{R\cdot T}\cdot\dot{v}_0$
Sum	$-\frac{p}{R\cdot T}\cdot\dot{V}_E$	$\frac{p}{R\cdot T}\cdot\dot{V}_E$	$-\frac{p}{R\cdot T}\cdot\dot{V}_E\cdot x$	$\frac{p}{R\cdot T}\cdot\left[\left(\dot{V}_E\cdot r+\frac{p_0}{p}\cdot\dot{v}_0\right)\cdot x_w+\dot{V}_E\cdot(1-r)\cdot x-\frac{p_0}{p}\cdot\dot{v}_0\right]$
Total	0		$\frac{p}{R\cdot T}\cdot\left(\dot{V}_E\cdot r+\frac{p_0}{p}\cdot\dot{\upsilon}_0\right)\cdot x_w-\frac{p_0}{R\cdot T}\cdot\dot{\upsilon}_0-\frac{p}{R\cdot T}\cdot\dot{V}_E\cdot r\cdot x$	

Where: x_w – molar fraction of oxygen in fresh respiratory gas, x – molar fraction of oxygen in inhaled breathing gas, x_0 – molar fraction of oxygen in breathing space before the diving apparatus is started, p – pressure at diving depth, p_0 – normal pressure, r – ventilated volume ratio of small bag to large bag $\mathrm{r}=\frac{\mathrm{u}}{\mathrm{U}-\mathrm{u}}$, R – universal gas constant, T – absolute temperature, U – volume of large bag, u – volume of small bag, $\dot{V}_E$ – ventilation of lungs, $\dot{\upsilon}_0$ – stream of consumed oxygen.

(Figure 4.2). After setting the values of streams, the ordinary first order differential equation can be written[6]:

$$\frac{p}{R\cdot T}\cdot(U-u)\cdot\frac{\partial x}{\partial t}=\frac{p}{R\cdot T}\cdot\left(\dot{V}_E\cdot r+\frac{p_0}{p}\cdot\dot{\upsilon}_0\right)\cdot x_w-\frac{p_0}{R\cdot T}\cdot\dot{\upsilon}_0-\frac{p}{R\cdot T}\cdot\dot{V}_E\cdot r\cdot x \quad (4.1)$$

where $\frac{\partial x}{\partial t}$ is elementary change in oxygen content ∂x in the large bag for elementary time change ∂t, and other designations as in Figure 4.2.

The solution of equation (4.1) in the form of mathematical proof is presented in Table 4.2:

$$x(t)=x_w-\varepsilon_p\cdot\varepsilon\cdot\varepsilon_k+\left(x_0-x_w+\varepsilon_p\cdot\varepsilon\cdot\varepsilon_k\right)\cdot exp\left(-\frac{\dot{\upsilon}_0\cdot r^2}{\varepsilon_0\cdot u}\cdot t\right) \quad (4.2)$$

where x_0 is molar fraction of oxygen in the breathing space before the diving apparatus is started, ε_k is *module*[7] representing the structural solution of the breathing space in the diving apparatus $\varepsilon_k=\frac{1-x_w}{r}$, ε is *breathing module*[8] $\varepsilon=\frac{\dot{\upsilon}}{\dot{V}_E}$, and ε_p is *pressure module* $\varepsilon_p=\frac{p_0}{p}$.

It follows from Equation (4.2) that oxygen content x in the large bag for time $t = 0$ from definition is $x(t=0)\stackrel{\text{def}}{=}x_0$ and for the time going to infinity $t\to\infty$ it is $x(t\to\infty)=x_s=x_w-\varepsilon_p\cdot\varepsilon\cdot\varepsilon_k$, where x_s means stable value $x_s(t_s)$ obtained after stabilization value t_s, which is no longer function of time t with assumed accuracy ϵ: $x_s(\epsilon)\neq f(t)$ for dive parameters $\varepsilon_p(t)$, $\varepsilon(t)=idem$ invariable in time t.

In Equation (4.2), the coefficient $\varepsilon_k=\frac{1-x_w}{r}$ is a formally dimensionless ratio[9] constituting as *criterion number*[10] describing the structural solution of the breathing space in the diving apparatus, whose value for a given solution is constant during the diving process $\varepsilon_k = idem \neq f(t)$, where time t refers to time of dive. It will be referred to here as structural module ε_k.

The ratio $\varepsilon_p=\frac{p_0}{p}$ is called here the pressure module, which is the *criterion number* of external conditions. It decreases with diving depth H: $H\nearrow\ \Rightarrow \varepsilon_p\searrow$. With the increase of depth H, the *pressure module* ε_p has less and less effect on the decrease in the stable oxygen content in the breathing gas, which is breathed by the diver x_s, due to its inverse proportionality to pressure p: $x_s\sim\varepsilon_p=\frac{p_0}{p}$. For the constant depth $H = idem$ the *pressure module* ε_p is constant value: $\varepsilon_p = idem \mid H = idem$.

The module $\varepsilon=\frac{\dot{\upsilon}}{\dot{V}_E}$ is called here the *breathing module*; for standard pressure p_0 it will be marked as ε_0, connecting oxygen consumption $\dot{\upsilon}_0$ and ventilation of lungs $\dot{V}_E$ with work W_0 done by a man on surface[11] $\dot{\upsilon}_0=f(W_0)$. The use of the *breathing module* related to normal conditions gives an opportunity to develop a draft design of a diving apparatus, because these values are determined with good accuracy, but in

Table 4.2

Derivation of the ventilation model related to oxygen content *x* in function of time *x* = *f*(*t*) for a semi-closed circuit rebreather SCR with a system of bags placed in above another

A:	$\dot{\upsilon}_0 \neq f(H)$		
T:	$x \rightarrow x = f(t)$		
P:	1°	$\frac{p}{R \cdot T} \cdot (U-u) \cdot \frac{\partial x}{\partial t} = \frac{p}{R \cdot T} \cdot \left(\dot{V}_E \cdot r + \frac{p_0}{p} \cdot \dot{\upsilon}_0 \right) \cdot x_W - \frac{p_0}{R \cdot T} \cdot \dot{\upsilon}_0 - \frac{p}{R \cdot T} \cdot \dot{V}_E \cdot r \cdot x$	from mole balance of oxygen
	2°	$\frac{\partial x}{\partial t} = \frac{\dot{V}_E \cdot r}{U-u} \cdot x_w - \frac{p_0}{p} \cdot \frac{\dot{\upsilon}_0}{U-u} \cdot (1-x_w) - \frac{\dot{V}_E \cdot r}{U-u} \cdot x$	ordinary first order differential equation
	3°	$\forall_{a=\frac{\dot{V}_E \cdot r}{U-u} \cdot x_W - \frac{p_0}{p} \cdot \frac{\dot{\upsilon}_0}{U-u}(1-x_W) \wedge b = \frac{\dot{V}_E \cdot r}{U-u}} \; dx = (a - b \cdot x) \cdot dt$	from 2°
	4°	$\frac{dx}{b \cdot x - a} + dt = 0$	from 3°
	5°	$\int \frac{dx}{b \cdot x - a} + \int dt = const = c$	from 4° indefinite integral definition
	6°	$\frac{1}{b} \cdot ln\lvert b \cdot x - a \rvert + t = c$	from 5°
	7°	$ln\lvert b \cdot x - a \rvert \equiv b \cdot (c - t) = c' - b \cdot t$	c′–new constant of integration
	8°	$e^{(c' - b \cdot t)} = c'' \cdot e^{-b \cdot t} \equiv b \cdot x - a$ c''– new constant of integration	from 7° and natural logarithm definition
	9°	$\forall_{t \rightarrow 0} \, x \rightarrow x_0 \Rightarrow c'' = b \cdot x_0 - a \Rightarrow (b \cdot x_0 - a) \cdot e^{-b \cdot t} = b \cdot x - a$ $\exists_{\frac{a}{b} = x_W - \varepsilon_p \cdot \varepsilon \cdot \varepsilon_k} \; x(t) = \left(x_0 - \frac{a}{b} \right) \cdot e^{-b \cdot t} + \frac{a}{b}$	from 8° i 3° $\varepsilon_k = \frac{1-x_w}{r}$ $\varepsilon_p = \frac{p_0}{p}$ $\varepsilon = \frac{\dot{\upsilon}}{\dot{V}_E}$
	10°	$x(t) = x_w - \varepsilon_p \cdot \varepsilon \cdot \varepsilon_k + (x_0 - x_w + \varepsilon_p \cdot \varepsilon \cdot \varepsilon_k) \cdot exp\left(-\frac{\dot{\upsilon}_0 \cdot r^2}{\varepsilon_0 \cdot u} \cdot t \right)$	from 2°, 8° i 9° q.e.d.

A – assumption, T – thesis, P – proof (evidence)

where: p – pressure at diving depth, p_0 – normal pressure, R – universal gas constant, T – absolute temperature, U – volume of the large bag, u – volume of the small bag, r – volume ratio of the small bag to the large bag $r = \frac{u}{U-u}$, $\dot{V}_E$ – ventilation of lungs, $\dot{\upsilon}_0$ – stream of consumed oxygen, x – molar fraction of oxygen inhaled breathing gas, x_w – molar fraction of oxygen in fresh breathing gas, x_0 – molar fraction of oxygen breathing space before diving apparatus is started, ε_k – module representing structural solution of breathing space in diving apparatus, ε – breathing module, ε_p – pressure module, $\frac{\partial x}{\partial t}$ – elementary change in oxygen content ∂x in large bag with elementary time change ∂t.

general it should be assumed that the *breathing module* can be dependent on the pressure $\varepsilon = f(p)$. This assumption must, however, be verified experimentally, as the workload under water W consists of additional types of physical work that do not occur on the surface.

For example, work done to overcome greater than in the air resistance associated with an increased pressure difference between the pulmonary centroid[12] and the center of buoyancy of the breathing space in the apparatus in immersion. Expending extra effort to overcome the resistance associated with the higher density of breathing gas ρ, causing an increase in flow resistance Δp posed by elements of the breathing space in the apparatus,[13] is also significant. In the breathing space of the diving apparatus there are also spatial obstructions that decrease the ventilation in some spaces, so-called *dead spaces*, whose impact on the safety of decompression should also be taken into account. Another condition that is difficult to determine unequivocally is an individual person-dependent phenomenon known as *bradycardia*,[14] associated with this slowed breathing f occurring in immersion, reducing ventilation of lungs $\dot{V}_E$ and, consequently, the breathing space in the diving apparatus (Birch MacLaren, & George, 2009). Hence, the assumption is that the *breathing module* ε should depend on pressure $\varepsilon = f(\varepsilon_p)$. It follows from the many years of experience that the *breathing module* ε_i is also an individualized function of *morphology*[15] of a diver i.

SPEED OF STABILIZATION

An important factor is the speed of the system's response to the stepwise forced change in the partial pressure of oxygen $p(t)$ in the circulating breathing gas.[16] For this purpose, the ventilation model (4.2) can be multiplied by pressure p present at diving depth H: $p(t) = p \cdot x(t)$. After ordering, the following can be written:

$$p(t) = p \cdot x_w - p_0 \cdot \varepsilon_0 \cdot \varepsilon_k + \left(p \cdot x_0 - p \cdot x_w + p_0 \cdot \varepsilon \cdot \varepsilon_k\right) \cdot exp\left(-\frac{\dot{\upsilon}_0 \cdot r^2}{\varepsilon \cdot u} \cdot t\right) \tag{4.3}$$

where $p(t)$ is partial pressure of oxygen in function of time t and p is pressure present at diving depth H.

Using the formula (4.3) it is possible to estimate stabilization times $t_s(\Delta p_{kr})$ needed to achieve stable value of oxygen partial pressure p_s in the breathing space of the diving apparatus with accuracy up to the specified critical pressure difference $\Delta p_{kr} = p(t) - p(t \to \infty)$. The stabilization times calculated in this way are close to stabilization times $t_s(\Delta p_{kr})$ in SCR with a nozzle constant dosing system (see previous chapters) (Kłos, 2007a; Kłos, 2007b; Kłos, 2019).

The theoretical process of oxygen partial pressure stabilization $p(t)$ in the breathing space of the SCR CRABE SCUBA is shown in Figure 4.3.

An example of real stabilization is shown in Figure 4.4.

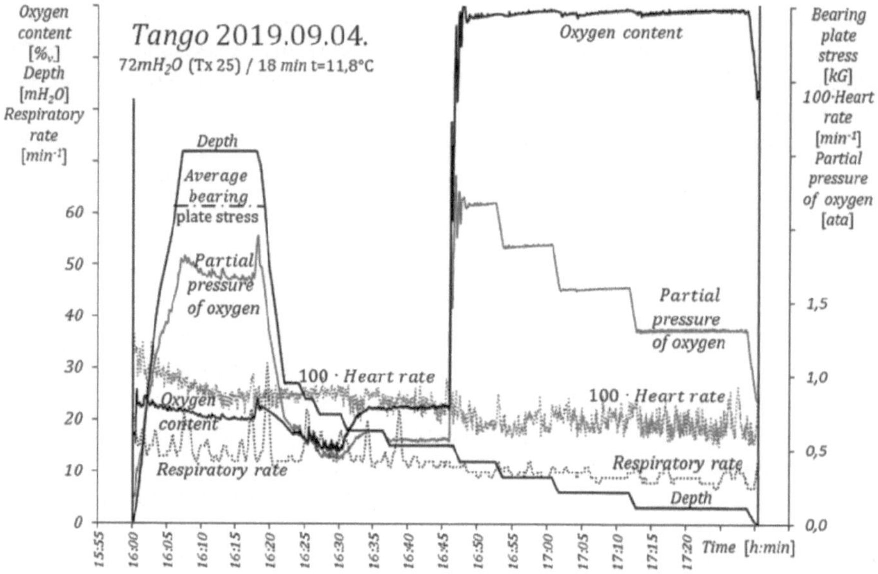

Figure 4.3 Theoretical times of partial pressure stabilization $p(t)$ of oxygen in the large bag.[17]

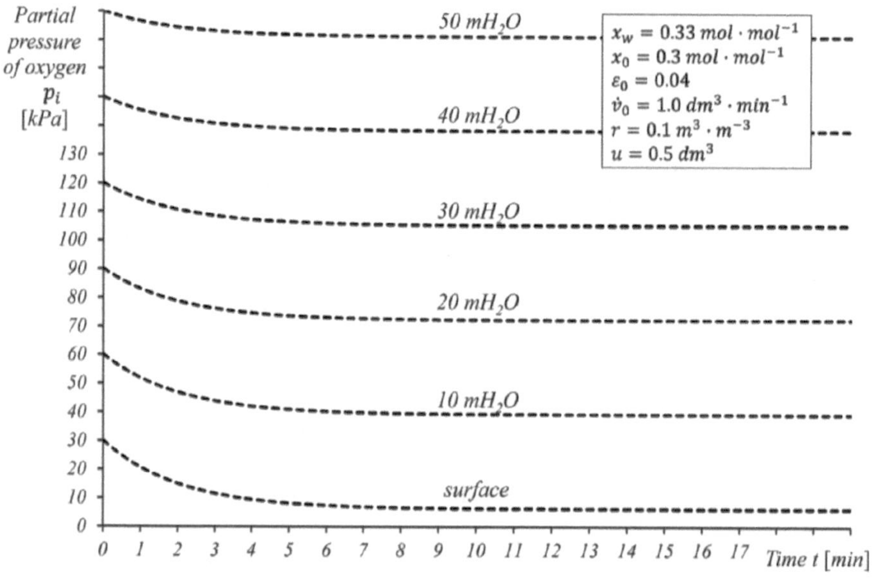

Figure 4.4 Example process of stabilizing the oxygen content $x(t)$ and its partial pressure $p(t)$ during a dive with CRABE apparatus and *premix* $25.0^{+0.5}_{-0.0}\%_v O_2 \vdots 35\%_v N_2 \vdots 40^{+1}_{-0}\%_v He$.

STABLE CONTENT

Equation $x(t \to \infty) = x_s = x_w - \varepsilon_p \cdot \varepsilon \cdot \varepsilon_k$ derived from an analysis of the model (4.2):

$$x_s = x_w - \frac{1-x_w}{r} \cdot \frac{\dot{\upsilon}}{\dot{V}_E} \cdot \frac{p_0}{p} = x_w - \varepsilon_k \cdot \varepsilon \cdot \varepsilon_p \Big|_{t \to \infty} \tag{4.4}$$

For the assumption $\varepsilon = idem$, dependence $x_s = f_{xw,\varepsilon,r}(\varepsilon_p)$ is equation of straight line $y = \alpha \cdot x + \beta$ with negative directional coefficient $\alpha = -\varepsilon_k \cdot \dot{\upsilon}$. Thus, for the domain of positive numbers it is a monotonically decreasing function. The transformed Equation (4.4) leads to dependence $x_w - x_s = f_{xw,\varepsilon,r}(p)$, which is the hyperbolic power function[18] decreasing together with depth H, always greater than zero $x_w - x_s = \varepsilon_k \cdot \varepsilon \cdot \frac{p_0}{p} > 0$, because $\varepsilon_k \geq 0 \wedge \varepsilon \geq 0$.

MODELING

Transforming boundary mathematical model (4.4), the following equation can be obtained: $\frac{p \cdot x_w - p_s}{p_0} = \frac{\Delta p}{p_0} = \frac{1-x_w}{r} \cdot \frac{\dot{\upsilon}}{\dot{V}_E}$. This equation can be further transformed into the form: $\frac{p \cdot x_w - p_s}{p_0} = \frac{\Delta p}{p_0} = \frac{1-x_w}{r} \cdot \frac{\dot{\upsilon}}{\dot{V}_E}$, where p represents the pressure at diving depth H, p_s is the actual partial pressure of oxygen in the breathing gas inhaled by the diver for determined structural conditions $\varepsilon_k = idem$ and diving conditions ε_p, $\varepsilon = idem$, and $\Delta p = p \cdot x_w - p_s$ is the difference in partial pressure of oxygen in the *premix* and the breathing gas inhaled by the diver. For different depths H, the *breathing module* $\varepsilon = \frac{\dot{\upsilon}}{\dot{V}}$ of oxygen consumption $\dot{\upsilon}$ for lung ventilation $\dot{V}_E$ is function of diving depth $\varepsilon = f(H)$:

$$\forall_{\varepsilon_p,\varepsilon,\varepsilon_k = idem}\; p \cdot x_w - p_s = \Delta p = \varepsilon \cdot \frac{1-x_w}{r} \cdot p_0 \tag{4.5}$$

where ε is the breathing module being the ratio of oxygen consumption to lung ventilation $\frac{\dot{\upsilon}}{\dot{V}_E}$ at diving depth H.

Interesting from the point of view of applications used for predicting safe decompression is a model combining total pressure p at diving depth H with oxygen partial pressure p_s, which can be obtained by multiplying stable content x_s from dependence (4.4) by hydrostatic pressure p at diving depth H:

$$p_s = x_w \cdot p - \varepsilon_k \cdot \varepsilon \cdot p_0 \tag{4.6}$$

For x_w, ε, $r = idem$, dependence (4.6) is equation of straight line $y = \alpha \cdot x + \beta$ having positive directional coefficient $\alpha = x_w$. Therefore, it should be a monotonically increasing function, with positive values for positive $\mathbb{R}_+$ arguments, which is confirmed by the results of our own research, as shown in Figure 4.4 (Kłos, 2016).

PARAMETERS OF THE MODEL

The diving apparatus *design module* $\varepsilon_k = \frac{1-x_w}{r}$ can be determined from direct measurement[19] of the volume ratio of the small bag and the large bag U: $r = \frac{u}{U-u}$ and oxygen content in premix x_w. Three SCR CRABE SCUBA were used in the research. The volume of breathing bags in these apparatuses was measured. The results are presented in Table 4.3.

The measurements were made using a vial having volume $V = (500 \pm 5)cm^3$. As it was necessary to use the vial six times to measure the maximum volume, the minimum systematic error of the volume measurement was not lower than $\Delta V > 30\ cm^3$. On the basis of the measurements, the volume of the small bag was assumed at the level of: $u = (0.50 \pm 0.01)dm^3$ and of the large sack $U = (5.30 \pm 0.03)dm^3$. Hence, the ratio of sack volumes can be assumed at the level[20] $r = (0.104 \pm 0.003)\frac{dm^3}{dm^3}$.

BREATHING MODULE

Figure 4.5 presents the results collected during the $N = 300$ series of dives carried out by a group of $l = 12$ divers and submitted for analysis by the manufacturer.

Figure 4.5 shows the minimum ε^{min} and the maximum values[21] of ε^{max} for the breathing modules as a function of the depth H estimated from the values provided by the manufacturer; additionally, preliminary results of own research are shown. Both these results and the results of our own research deviate from the values $\varepsilon_0 \in [0.035; 0.055]$ found in the literature concerning atmospheric pressure p_0 (PN-EN-14143, 2005). In Figure 4.5, slight dependence of the breathing module ε on pressure p can be seen: $\varepsilon = f(p)$. As shown in the analysis of the sensitivity of the model, the initial adoption of the mean value for the *breathing module* as $\bar{\varepsilon} \cong 0.065$ seems sufficiently accurate to be used for the initial modeling of ventilation of the breathing space in a diving apparatus.

Figure 4.4 presents an example of the experimental diving result for *premix* $x_w = 0.25\ Tx$ (Kłos, 2016). The dives were used to determine the stable oxygen content x_s and total pressure p, constituting the basis for calculating stable value of oxygen partial pressure p_s. Using the known values of the *structural module* $\varepsilon_k = \frac{1-x_w}{r}$,

Table 4.3
Results of direct measurements of the volume of breathing bags for three CRABEs

Particulars	Volume [dm^3]		
	1	2	3
Small bag u	0.50 ± 0.01	0.50 ± 0.01	0.50 ± 0.01
Large bag U	5.30 ± 0.03	5.35 ± 0.03	5.35 ± 0.03
Space between small bag and large bag $U - u$	4.90 ± 0.03	4.85 ± 0.03	4.90 ± 0.03

Vial was used to make measurement $V = (500 \pm 5)cm^3$

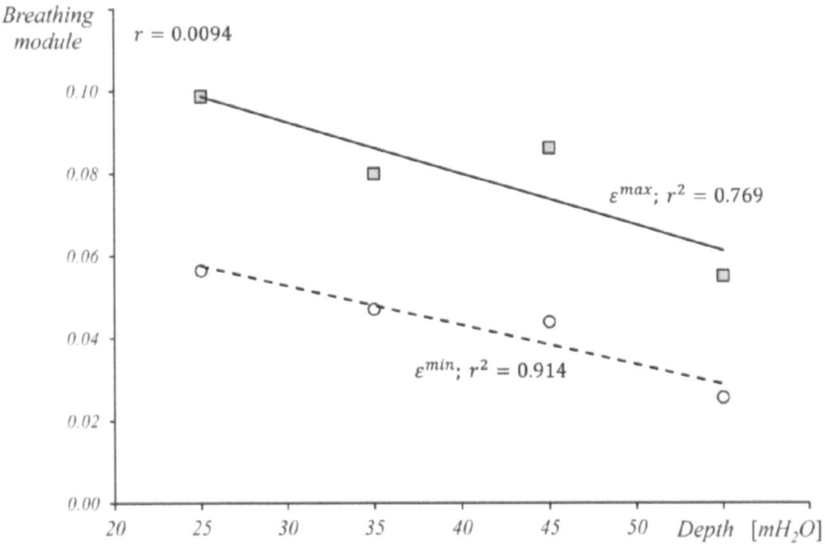

Figure 4.5 Practical values of minimum changes ε^{min} and maximum changes ε^{max} of the *breathing module* in the depth function calculated on the basis of the data $N = 300$ dives

the *pressure module* $\varepsilon_p = \frac{p_0}{p}$, the mean value of the breathing module $\bar{\varepsilon} \cong 0.065$, and using the dependence (4.6), the theoretical dependence can be drawn of the stable oxygen partial pressure p_s in the function of pressure p present at the maximum diving depth H. It should be noted, however, that the value of total pressure p recorded during the experiments was the pressure exerted by the atmosphere of the hyperbaric chamber, and the diving apparatus was additionally placed in the pool at the depth of approx. $h \cong 0.5\ mH_2O$. A satisfactory compatibility can be achieved between the measurement results and the theoretical model of ventilation (4.6) when recorded total pressure p is increased by the value of $p_+ = 5\ kPa$ associated with immersion of the apparatus below the surface of water in the pool (Figure 4.6).

The results of measurements were used to estimate the volume ratio of the bags in the apparatus: $r = (0.104 \pm 0.003) m^3 \cdot m^{-3}$. The value of the structural module ε_k during the tests with *Nitrox* for the oxygen content in the *premix* $x_w \cong 0.325\ m^3 \cdot m^{-3}$ did not change $\varepsilon_k = idem$ was $\varepsilon_k = \frac{1 - x_w}{r} \cong \frac{1 - 0.325}{0.104} \cong 6.49$. Taking into account the above values $\bar{\varepsilon} \cong 0.065$, $x_w \cong 0.325\ m^3 \cdot m^{-3}$ and $\varepsilon_k = 6.49$, a theoretical model defining the dependence of the stable partial pressure of oxygen p_s *vs* the total pressure p of the water environment at depth H: $p_s \cong 0.325 \cdot p - 0.454 \cdot p_0$ can be determined (Figure 4.6). The recorded stable values of oxygen partial pressures p_s for experimental dives from different years were $p_s \cong 0.334 \cdot p - 46.761$ showing high value of the determination coefficient $R^2 \cong 0.9785$, relatively rarely recorded in biological studies. It can be assumed that a good agreement between the theoretical and experimental results is obtained, despite the fact that the amount of work done by the divers was

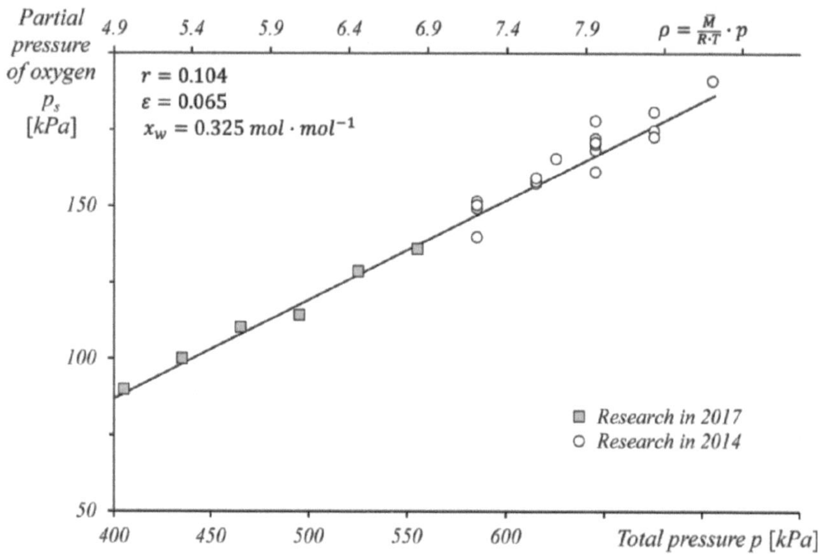

Figure 4.6 Practical values of stabilized partial pressure of oxygen p_s in the function of pressure p present at the depth H and density ρ of breathing gas $p_s = f(\rho)$, based on our own preliminary research carried out in 2014 and 2017; where ρ – density of the breathing gas $\rho = f(p)$, $\bar{M}$ – mean molar mass of the breathing gas, R – universal gas constant, T – thermodynamic temperature

not constant $W \neq idem$; hence for the investigations, the effect of the pressure applied to the underwater ergometer plate F on the stable oxygen content p_s did not seem significant (Figure 4.7).

SENSITIVITY ANALYSIS

From the measurements carried out, the volume ratio of the breathing bags was estimated as $\left(0.104 \pm 0.003\right)\frac{dm^3}{dm^3}$. The chromatographic method for measuring the oxygen content guarantees the maximum level of absolute error of $\Delta x_w = \pm 0.002\ m^3 \cdot m^{-3}$. Based on the results of the study carried out in 2014, the approximate dependence of the *breathing module* on pressure p was determined: $\varepsilon(p) \cong 0.074 - 2.7 \cdot 10^{-5} \cdot p$. For depth range $H \in 0\text{–}60\ mH_2O$ the variability in the *breathing module* value was within the range $\varepsilon \in 0.06\text{–}0.07$, hence it was assumed: $\varepsilon \cong 0.065 \pm 0.004$.

Based on the variability: oxygen content in premix x_w, of the breathing module ε, of volume ratio of bags r, of pressure range p at maximum diving depth H, the sensitivity of the function (4.6) was estimated for the specified tolerance ranges:

$$
\begin{aligned}
x_w &= 0.325 \qquad & \Delta x_w &= \pm 0.002 \\
r &= 0.10 \qquad & \Delta r &= \pm 0.01 \\
\varepsilon &= 0.065 \qquad & \Delta \varepsilon &= \pm 0.01
\end{aligned}
$$

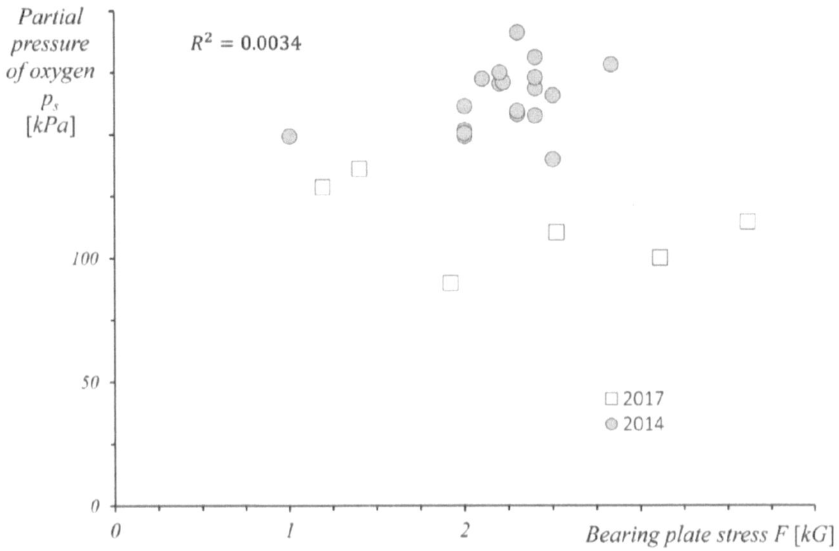

Figure 4.7 Results of measurements of the bearing stress F applied to the underwater ergometer plate for the same dives as in Figure 4.6.

Change in the stable value of oxygen partial pressure Δp_s will be a complete differential with regard to the adopted parameters x_w, ε, r, and pressure p, multiplied by their variability ranges:

$$\Delta p_s = \left|\frac{\partial p_s}{\partial \varepsilon}\right| \cdot \Delta\varepsilon + \left|\frac{\partial p_s}{\partial x_w}\right| \cdot \Delta x_w + \left|\frac{\partial p_s}{\partial r}\right| \cdot \Delta r + \left|\frac{\partial p_s}{\partial p}\right| \cdot \Delta p$$

$$\Delta p_s(\Delta\varepsilon) = \left|\frac{\partial p_s}{\partial \varepsilon}\right| \cdot \Delta\varepsilon = \left|\frac{1-x_w}{r} \cdot p_0\right| \cdot \Delta\varepsilon = \frac{1-0.325}{0.10} \cdot 100 \cdot (\pm 0.01) \cong \pm 6.8\, kPa$$

$$\Delta p_s(\Delta x_w) = \left|\frac{\partial p_s}{\partial x_w}\right| \cdot \Delta x_w = \left|p - \frac{\varepsilon \cdot p_0}{r}\right| \cdot \Delta x_w = \left(600 + \frac{0.065 \cdot 100}{0,10}\right) \cdot (\pm 0.002) \cong \pm 1.3\, kPa$$

$$\Delta p_s(\Delta r) = \left|\frac{\partial p_s}{\partial r}\right| \cdot \Delta r = \left|\frac{1-x_w}{r^2} \cdot p_0 \cdot \varepsilon\right| \cdot \Delta r = \frac{1-0.325}{0.10^2} \cdot 100 \cdot 0,01 \cdot (\pm 0.01) \cong \pm 0.7\, kPa$$

$$\Delta p_s(\Delta p) = \left|\frac{\partial p_s}{\partial p}\right| \cdot \Delta p = |x_w| \cdot \Delta p = 0.325 \cdot \Delta p \rightarrow \Delta p_s(\Delta p) \in (0;162)\, kPa$$

It follows from the calculations that for depth from $H = 20\ mH_2O$ the effect of changes in pressure Δp to diving depth H exceeds by more than sevenfold the combined effects of oxygen partial pressure stable of fluctuation Δp_s caused by changes in oxygen content in *premix* Δx_w, the *breathing module* $\Delta\varepsilon$, and volume ratio of bags Δr. However, at small depths $H < 20\ mH_2O$ the effect of fluctuation in the *breathing module* $\Delta\varepsilon$ is comparable to the effect of pressure changes Δp. To stabilize the partial pressure of oxygen in the breathing space p_s at small depths $H < 20\ mH_2O$, fresh *premix* can be supplied by the constant dosing system switching at depth of not less than $H < 20\ mH_2O$. Activating this supply above depth $H \geq 25\ mH_2O$ should not have any practical significance, because at depths above $H = 25\ mH_2O$ the effect of the pressure present at the diving depth exceeds the effect of the aforementioned fluctuations and considered here magnitudes by an order of magnitude.

APPARATUS DESIGN PROCESS

From the calculations of the sensitivity of the model (4.6), it can be seen that stable partial pressure of oxygen p_s strongly depends on depth $H = f(p)$. For depths greater than $H > 20\ mH_2O$ the effect of anticipated changes in oxygen content in *premix* x_w, the breathing module ε or volume ratio of bags r is practically negligible. The approximation of the value of the *breathing module* with its range of values determined for normobaric conditions $\varepsilon \cong \varepsilon_0$ may be sufficient for the initial design process of the diving apparatus, because as this is shown by the practical dependence $\varepsilon(p) \cong 0.074 - 2.7 \cdot 10^{-5} \cdot p$ obtained from the measurements, for depths greater than $H > 20\ mH_2O$ the value of the *breathing module* ε goes to the values recorded on the surface $\varepsilon \to \varepsilon_0$.

The apparatus for *Nitrox* mixtures *Nx* is most often designed for limited depth range $H \in 0\text{–}50\ mH_2O$. On surface $H = 0\ mH_2O$ the stable partial oxygen pressure should not fall below the physiological level $p_s > 20\ kPa$, and for the maximum depth $H = 50\ mH_2O$ most often it is assumed that it should not exceed $p_s \leq 160\ kPa$, due to toxic effects of oxygen. Starting from these assumptions, it can initially be calculated that for surface $H = 0\ mH_2O$, the oxygen content in *premix* x_w should not be less than $x_w > \frac{p_s}{p} = \frac{20\,kPa}{100\,kPa} = 0.2\frac{mol}{mol}$. For depth $H = 50\ mH_2O$ the oxygen content in *premix* x_w should not exceed $x_w < \frac{p_s}{p} = \frac{160\,kPa}{600\,kPa} \cong 0.27\frac{mol}{mol}$. For short excursion times it is possible to accept the value of stable partial pressure of oxygen at level $p_s \leq 200\ kPa$ (Kłos, 2012). Hence, the upper oxygen content in the *premix* may theoretically be $x_w < \frac{p_s}{p} = \frac{200\,kPa}{600\,kPa} \cong 0.33\frac{mol}{mol}$. In case of failure and when dealing such a situation by ventilating the breathing space of the apparatus, the value greater than $x_w \cong 0.33\frac{mol}{mol}$ is also acceptable, as it is not possible to fully ventilate the breathing space of the apparatus. The analysis shows that *premix Nx* 0.325 can be adopted for recommendation. Pressure drop $\Delta p = p \cdot x_w - p_s$ representing the difference between partial pressure of oxygen in *premix* $p_0 \cdot x_w$ and breathing gas inhaled by the diver p_s can be estimated for the surface at the level: $\Delta p = p_0 \cdot x_w - p_s = 100\,kPa \cdot 0.325\frac{mol}{mol} - 20\,kPa = 12.5\,kPa$ and for depth $H = 50\ mH_2O$ na: $\Delta p = p \cdot x_w - p_s = 600\,kPa \cdot 0.325\frac{mol}{mol} - 20\,kPa = 175\,kPa$. In line

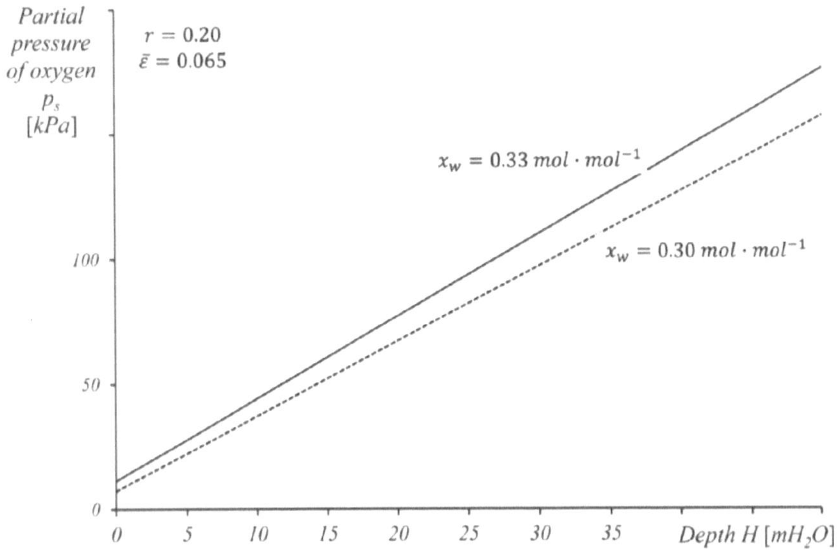

Figure 4.8 Example of dependence of stabilized partial pressure of oxygen $p_s = f_{x_w,\bar{\varepsilon},r}(H)$ in function of depth H for the *design module* $\varepsilon_k = \frac{1-x_w}{r}$ and mean value $\bar{\varepsilon}$ of the *breathing module* $\varepsilon = \frac{\dot{\upsilon}}{\dot{V}_E}$.

with the dependence (4.5), the recommended volume ratio of bags r to the normobaric conditions $H = 0\ mH_2O$, where $p = p_0$ i $\varepsilon \equiv \varepsilon_0$: $r = \varepsilon_0 \cdot \frac{1-x_w}{\Delta p} \cdot p_0 = \frac{1}{25}\frac{m^3}{m^3} \cdot \frac{1-0.33}{13}\frac{mol}{mol \cdot kPa} \cdot 100\,kPa \cong 0.2\frac{m^3}{m^3}$ can be precalculated. The dependence of stable partial pressure of oxygen p_s in function of depth H for the mean value of the *breathing module*[22] $\bar{\varepsilon} = 0.065$ in depth range $H \in 0\text{–}50\ mH_2O$ and volume ratio of bags $r = 0.2$ are shown in Figure 4.8.

In order to reduce the value of stable partial pressure of oxygen p_s at maximum depth $H = 50\ mH_2O$, the manufacturer applied a different technical solution for the apparatus, adopting the volume ratio of bags as $r = 0.1\frac{m^3}{m^3}$. For such a structure and *premix* with oxygen content $x_w = 0.33\frac{mol}{mol}$ there will be a decrease in stable partial pressure of oxygen on the surface $p_s(p = p_0)$ to the hypoxic level $p_s < 20\ kPa$. Only for *premix* with oxygen content $x_w \cong 0.52\frac{mol}{mol}$ partial pressure of oxygen p_s will increase to the value $p_s \cong 18\ kPa$ acceptable under emergency conditions. The manufacturer has additionally introduced a constant *premix* dosing system to support apparatus ventilation, which switches on within depth range $H \in 22\text{–}24\ mH_2O$. The range within which the ejector support is switched on/off during the dive within the depth range $H \in 22\text{–}24\ mH_2O$ is marked in Figure 4.9.

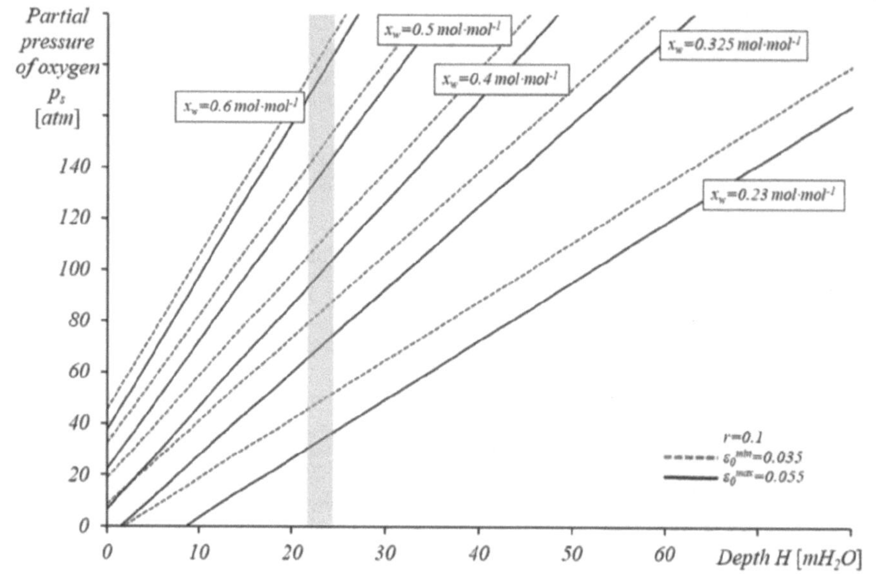

Figure 4.9 Examples[23] of theoretical ranges of stable oxygen partial pressures p_s in inhaled breathing gas in function of function of depth H and various premixes with oxygen content x_w as used by the manufacturer of the apparatus.[24] For *premixes* with oxygen content $x_w \leq 0.5\ mol \cdot mol^{-1}$ for depths within the range $H \in$ 21–24 mH_2O, ejector support for dosing of fresh *premix* is used.

PREMIX

The analysis of the model (4.6) shows that for the SCR CRABE SCUBA, the minimal, stable partial pressure of oxygen p_s^{min} in the breathing gas inhaled by the diver will occur for the case of minimum pressure[25] p^{min} at minimum diving depth H^{min} and maximum oxygen consumption $\dot{v}^{max}$, entailing maximization of the *breathing module* ε^{max}. The maximum stable partial pressure of oxygen p_s^{max} in the breathing gas inhaled by the diver will occur for the maximum pressure during the dive p^{max} and the minimum value of the breathing module ε^{min}. Thus, a system of equations:

$$\begin{cases} p_s^{max} = x_w \cdot p^{max} - \varepsilon_k \cdot \varepsilon^{min} \cdot p_0 \\ p_s^{min} = x_w \cdot p^{min} - \varepsilon_k \cdot \varepsilon^{max} \cdot p_0 \end{cases}$$

can be written, which for $p^{min} = p^{max} = p$ and approximation $\varepsilon \in$ 0.06–0.07 takes the form:

$$\begin{cases} p_s^{max} = x_w \cdot p - \varepsilon_k \cdot \varepsilon^{min} \cdot p_0 \\ p_s^{min} = x_w \cdot p - \varepsilon_k \cdot \varepsilon^{max} \cdot p_0 \end{cases} \tag{4.7}$$

For mixture Nx 0.325 and for volume ratio of bags $r = \frac{u}{U-u} = 0.104$, the structural module $\varepsilon_k = \frac{1-x_w}{r}$ will be $\varepsilon_k = 6.49$. Hence, for atmospheric pressure $p = p_0 \cong 100\ kPa$, the stable value of pressure p_s will be theoretically lower than or equal to $p_s \lesssim 13\ kPa$. For depth $H = 50\ mH_2O$, the corresponding pressure will be $p \cong 600\ kPa$ and the stable value of pressure p_s will theoretically be within $p_s \in 149–156\ kPa$. By performing calculations in a similar way, the ranges of stable partial oxygen pressures p_s can be predicted for various breathing gases (Figure 4.9). A system of breathing bags placed one above the other does not protect the diver against hypoxia at small depths for mixtures.

SUMMARY

The performed analyzes allowed for the development of a deterministic model of ventilation of the breathing apparatus space (4.2): $x(t) = x_w - \varepsilon_p \cdot \varepsilon \cdot \varepsilon_k + (x_0 - x_w + \varepsilon_p \cdot \varepsilon \cdot \varepsilon_k) \cdot exp\left(-\frac{\dot{\upsilon}_0 \cdot r^2}{\varepsilon \cdot u} \cdot t\right)$, where t is time, x is molar fraction of oxygen in inhaled breathing gas, x_w is molar mole oxygen in fresh respiratory agent, and x_0 is molar fraction of oxygen in the breathing space before starting the diving apparatus depending on the criterion numbers:

- $\varepsilon_k = \frac{1-x_w}{r}$ is the module representing the constructional solution of the breathing space of a diving apparatus
- $\varepsilon = \frac{\dot{\upsilon}}{\dot{V}_E}$ is the breathing module
- $\varepsilon_p = \frac{p_0}{p}$ is the pressure module

where p is pressure at diving depth, p_0 is normal pressure, U is volume of large bag, u is volume of small bag, r is volume ratio of small bag to large bag $r = \frac{u}{U-u}$, $\dot{V}_E$ is lung ventilation, and $\dot{\upsilon}$ is stream oxygen being used.

The aim of the preliminary research carried out was to determine parameters for a breathing space ventilation model in a diving apparatus by comparing the results of measurements of a stable value of oxygen partial pressure p_s with the theoretical values derived from the model (4.2). The key parameter of this model is the breathing module ε related to the working pressure p_H: $\varepsilon = f(\varepsilon_p)$.

NOTES

1 Of a predetermined quantitative composition, constant during the process of diving.
2 During oxygen decompression.
3 The term *premix* was used as the fresh breathing gas contained in gas cylinders is a gas mixture of unchanged composition.
4 Inspired/expired steam of volume V_E.
5 moles

6 In differential equations, the function occurring under the derivative designation is unknown. If the unknown is a function of one variable, then the equation is called *ordinary*. The order of the highest derivative in the equation is the *order of the equation*.

7 Here the *module* is referred to as a formally dimensionless quantity; the modules will also be referred to as *criterion numbers*.

8 Ventilatory equivalent for oxygen $V_E(O_2)$.

9 Sometimes the formally dimensionless ratio is dependent on the values on the basis of which it was obtained, which is why their designation is sometimes given; for example, for real gases, molar fraction x_n having dimension $\left[x_n\right]=\frac{mol}{mol}$ is not equal to volume fraction x_v having the dimension $\left[x_v\right]=\frac{m^3}{m^3}$: $x_n \neq x_v$ with regard to the numerical value; therefore, often the designations of the criterion numbers are also marked

10 As previously noted, dimensionless relations will be called *modules* or *criterion numbers*, except the breathing module ε, which is commonly referred to as the *breathing module*.

11 Related to the measured values of load W_0 on the surface.

12 Centroid is understood here as a geometric center of the space lying in the geometric center of its shape equivalent to the center of buoyancy.

13 Usually in investigations on oxygen consumption dependent on effort, resistance of the breathing space in a measuring apparatus was negligible in relation to breathing resistance generated by a diving apparatus already on the surface.

14 Slowed heart rate; in immersion, the pressure on limbs increases and there occurs a contraction of skin vessels due to cold, causing an increase in venous return, which causes an increase in stroke volume, which leads to a reduced heart rate.

15 Understood here as a science of the form, shaping, and constitution of organisms.

16 Speed of stabilization.

17 The results of theoretical calculations can be compared with the real stabilization process; see an example in Figure 4.4 (Kłos., 2016).

18 As $x_s \sim p^{-1}$.

19 For example, by flooding with water.

20 $\Delta r=\left|\frac{\partial r}{\partial u}\right|\cdot\Delta u+\left|\frac{\partial r}{\partial\left(U-u\right)}\right|\cdot\Delta\left(U-u\right)=\frac{1}{U-u}\cdot\Delta u+\frac{u}{\left(U-u\right)^2}\tau\cdot\Delta\left(U-u\right)=\frac{1}{4.8}\cdot 0.01+\frac{0.5}{4.8^2}\cdot 0.02\cong 0.002$

21 Value of *breathing module* $\varepsilon=f(p)$ can be estimated by transforming dependence (4.6):

$$\varepsilon=\frac{x_w\cdot p-p_s}{\varepsilon_k\cdot p_0}\cdot mol^{-1}$$

22 $\varepsilon(p)\cong 0.082-2.7\cdot 10^{-5}\cdot p \rightarrow \varepsilon(p=100\ kPa)=0.08 \wedge \varepsilon(p=600\ kPa)=0.06$.

23 *premix* having oxygen content x_w=0.23 $mol\cdot mol^{-1}$ is *Trimix* whereas the other ones are *Nitrox* mixtures.

24 When the values of the breathing module determined for normal pressure are applied, $\varepsilon_0 \in [0.035; 0.05]$ are applied.

25 Diving depth.

REFERENCES

Birch K., MacLaren D., & George K. 2009. *Fizjologia sportu-krótkie wykłady*. Warszawa: Wydawnictwo Naukowe PWN. ISBN 978-83-01-15460-8.

Kłos R. 2000. *Aparaty Nurkowe z regeneracją czynnika oddechowego*. Poznań: COOPgraf. ISBN 83-909187-2-2.

Kłos R. 2002. Mathematical modelling of the breathing space ventilation for semi-closed circuit diving apparatus. *Biocybernetics and Biomedical Engineering* 22, 79–94.

Kłos R. 2007a. Modelowanie procesów wentylancji obiektów normo- i hiperbarycznych. Higher PhD diss., The Polish Naval Academy. PL ISSN 0860-889X nr 160A.

Kłos R. 2007b. *Mathematical modelling of the normobaric and hyperbaric facilities ventilation*. Gdynia: Polish Hyperbaric Medicine and Technology Society. ISBN 978-83-924989-0-2.

Kłos R. 2011. *Możliwości doboru dekompresji dla aparatu nurkowego typu CRABE*. Gdynia: Polish Hyperbaric Medicine and Technology Society. ISBN 978-83-924989-4-0.

Kłos R. 2012. *Możliwości doboru ekspozycji tlenowo-nitroksowych dla aparatu nurkowego typu AMPHORA – założenia do nurkowań standárdowych i eksperymentalnych*. Gdynia: Polish Hyperbaric Medicine and Technology Society. ISBN 978-83-924989-8-8.

Kłos R. 2016. *System trymiksowej dekompresji dla aparatu nurkowego typu CRABE*. Gdynia: Polish Hyperbaric Medicine and Technology Society. ISBN 978-83-938322-5-5.

Kłos R. 2019. Modelling of the normobaric and hyperbaric facilities ventilation. *International Journal of Mechanical Engineering and Applications* 7, 26-33. DOI: 10.11648/j.ijmea.20190701.14.

5 Tests of Construction with a Metering Bellows Dispenser

The basic parameters of breathing equipment tested in the laboratory, without the participation of experimental divers, are:

- respiratory parameters[1]
- stability parameters[2]
- ergonomic parameters[3]
- construction and functional parameters[4]
- strength and reliability parameters[5]
- operating parameters of the most important components[6]

In the laboratory tests conducted without participation of people, a breathing action simulator coupled with a modernized metabolic simulator was used (Kłos, 2002; Kłos, 2007). Such a test stand should make it possible to carry out into investigations on a ventilation model used in diving apparatus. The ventilation model of the diving apparatus provides data for the decompression model that enables planning adequate decompression. So far, investigations have focused on the combined ventilation and decompression model (Kłos, 2011; Kłos, 2016). At present, due to the fact that sufficiently large experimental data have been collected, it has been possible to separate these models and conduct independent studies on their adequacy and credibility.

METABOLIC SIMULATOR

Implementation of newly developed structural solutions applied in UBA is inseparable from carrying out experiments involving people who are exposed to risks associated with the inherent necessity to test new or modernized diving technologies.

The basis for modeling diving technologies is checking decompression distributions. For modeling decomposition distributions, it is important to check changes in the breathing module ε in function of depth: $\varepsilon = f(H)$. On this basis, the accuracy of ventilation model of the breathing space in the diving apparatus can be determined. The ventilation model can be used to determine stable oxygen content for individual depths: $x_s = f(H)$. When the fluctuation of stable oxygen content x_s is taken into account for individual depth H, this model contributes to planning adequate decompression together with the *central nervous syndrome* CNSyn model, determining the risk of oxygen toxicity in planned exposures (Kłos, 2012). By grouping the premises derived from the ventilation and CNSyn models, and using the decompression model, exposure distributions can be determined in the form of decompression tables that are verified through experiments with participation of human beings.

Use of a breathing simulation stand together with metabolic gas exchange in hyperbaric conditions can reduce the risk of big errors in the initial phase of investigations by providing answers to a number of important problems related to the stability of composition of a breathing gas in the breathing space of a diving apparatus before experiments involving humans are commenced.

Simulation of gas exchange in the process of breathing in hyperbaric conditions, in addition to mechanical simulation of the breathing cycle, should allow for intake of oxygen O_2 from the breathing space of the UBA and emission of carbon dioxide CO_2 and water vapor H_2O in the required ratio with the required accuracy and precision. A modernized metabolic simulator coupled with a breathing simulation stand and a hyperbaric simulator can be used for the study and validation of mathematical models of the ventilation process in the UBA.

In simple metabolic simulators, controlled dosing[7] of carbon dioxide CO_2 is used, disregarding the resultant increase in gas mass in the hyperbaric breathing space (Kyriazi, 1986). Such a procedure is justified when mass stream $\dot{m}_{CO_2}$ of carbon dioxide CO_2 is so small in relation to mass flow of the ventilation medium $\dot{m}$ and the whole mass in the ventilated breathing space $\dot{m}_0$ exchange that the carbon dioxide CO_2 metering $\dot{m}_{CO_2}$ does not disturb the ventilation process[8] to a significant extent.

Oxygen can be removed by catalytic oxidation of various chemical compounds[9] added to the breathing space. Such oxygen intake may be accompanied by simultaneous emission of carbon dioxide CO_2 and water H_2O (Loncar, 1992; Kłos, 2002; Frånberg, 2015). This is currently one of the best methods for simulating gas exchange.

As it has been mentioned, the mapping of the breathing process in hyperbaric conditions is connected not only with the possibility to carry out investigations under increased pressure and mechanical simulation of lung ventilation but also with mapping of gas exchange. These processes should be reproduced with reproducibility, accuracy, and precision required by the scientific research method. However, they cannot simulate all the details of the real human–machine ergonomic interaction, offering only a simplified model but adequate to the research project undertaken. In the case of simulation of gas exchange taking place in the body, the most important is oxygen consumption of $\dot{v}_{O_2}$ and the emission of carbon dioxide $\dot{v}_{CO_2}$ in its place. The most common are the following two values of the adopted *respiratory quotient*[10] $\varepsilon_{RQ} \in 0.75\text{–}0.8\ dm^3 \cdot dm^{-3}$ indicating the volume ratio $\varepsilon_{RQ} = \frac{\dot{v}_{CO_2}}{\dot{v}_{O_2}}$ at which emission $\dot{v}_{CO_2}$ of carbon dioxide CO_2 in place of consumed $\dot{v}_{O_2}$ of oxygen O_2 occurs. To simulate this process, catalytic oxidation of *ethanal*[11] CH_3CHO or *propanone*[12] $(CH_3)_2CO$ can be used (Kłos, 2002):

$$2 \cdot CH_3CHO + 5 \cdot O_2 \xrightarrow{catalyst} 4 \cdot CO_2 + 4 \cdot H_2O \; \varepsilon_{RQ} = 0.80$$
$$\left(CH_3\right)_2 CO + 4 \cdot O_2 \xrightarrow{catalyst} 3 \cdot CO_2 + 3 \cdot H_2O \; \varepsilon_{RQ} = 0.75 \quad (5.1)$$

where ε_{RQ} is *respiratory quotient.*

The catalyst in reaction (5.1) can be colloidal platinum *Pt* on a ceramic carrier, diatomaceous earth, alumina, etc. The reaction takes place in the volume of the breathing gas drawn in by the breathing simulator in the reactor, and then the gas is returned together with combustion products to the breathing space in the UBA placed in the hyperbaric chamber.

In reaction (5.1) the volume of liquid substrates is so small in relation to the volume of the gaseous products formed and the entire breathing space of the UBA that most often it does not cause any serious disturbances, significant for the balance of the volume of the breathing space. In addition, the gas is humidified, which makes the simulation process more real and creates good chemisorption conditions on the filling of the carbon dioxide CO_2 absorber used in the SCR (Kłos, 2009). The significant exothermic effect of the oxidation process may generate some problems, hence the reaction products should be cooled once they leave the reactor, before the breathing gas returns to the breathing space of the SCR.

Implementation-oriented studies on the currently used metabolic simulator were presented earlier in Chapter 2. The simulation stand consisting of breathing, metabolic, and aqueous hyperbaric environment simulators is shown in Figure 5.1. The previous one was described earlier (Kłos, 2002; Kłos, 2007).

The catalytic oxidation of *propanone*[13] was selected for simulation of metabolic processes. *Propanone* $(CH_3)_2CO$ was metered to the reactor by means of SMART Digital S pump *DDA made by Danish company Grundfos Holding A/S*, which offers dosing with accuracy required for the counterpressure occurring during the investigations. The reactor was filled with $0.5\%_m Pt$ catalyst type GA-20Pt produced by Katalizator sp. z o.o. on the Institute of Catalysis and Surface Chemistry (Polish Academy of Sciences in Krakow) license (Kłos, 2015).

FIGURE 5.1 Stand of combined simulators: respiratory, metabolic, and hyperbaric.

THE RESULTS OF THE SIMULATION-BASED INVESTIGATIONS

The research was based on volumetric measurements, which for a liquid such as *propanone* Me_2CO are dependent on temperature t and for oxygen O_2 also from pressure p. Most often the measurements were made at a temperature of approx. $t \cong 25°C$. During the measurements, constant pressure value $p = idem$ was maintained. Oxygen consumption calculations were referenced to the temperature and pressure present at the level of inhalation from the mouthpiece of the dive apparatus $\{t,p\}$, and referenced to standard conditions $T_0 \cong 273.15\ K$; $p_0 \cong 101.325\ kPa$. For each measurement, the molar volume of oxygen was determined depending on the temperature present in the respiratory circuit of the diving apparatus.

Propanone is characterized by molar mass $M_{Me_2CO} \cong 58.08\ kg \cdot kmol^{-1}$, density $d_{20°C} \cong 0.791\ kg \cdot dm^{-3}$, melting point $t_t \cong -95°C$ and boiling point $t_w \cong 56°C$. It follows from reaction (5.1) that 1 *mol* of Me_2CO reacts with 4 moles of oxygen O_2, hence mass stream of consumed oxygen $\dot{m}_i$ is proportional to molar mass of oxygen M_i, molar mass of *propanone* M_{Me_2CO}, volume flow of *propanone* $\dot{V}_{Me_2CO}$ and Me_2CO density $d_{25°C}$: $\dot{m}_i = 4 \cdot \frac{M_i}{M_{Me_2CO}} \cdot \dot{V}_{Me_2CO} \cdot d_{25°C}$. Hence, mole stream of oxygen $\dot{n}_i = \frac{\dot{m}_i}{M_i} = 4 \cdot \frac{1}{M_{Me_2CO}} \cdot \dot{V}_{Me_2CO} \cdot d_{25°C}$. Thus the stream of consumed oxygen referred to the normal conditions $\dot{v}_0$ will be $\dot{v}_0 = v_0 \cdot \dot{n}_i = 4 \cdot \frac{v_0}{M_{Me_2CO}} \cdot \dot{V}_{Me_2CO} \cdot d_{25°C}$, where molar volume of oxygen under normal conditions $v_0 \cong 22.39\ dm^3 \cdot mol^{-1}$. Ventilation flow referred to standard conditions $\dot{V}_E$ will be proportional to cylinder capacity V_E of the piston breathing device, its frequency $\dot{n}$, pressure conditions p, and measurement temperature p: $\dot{V}_E = V_E \cdot \dot{n} \cdot \frac{p}{p_0} \cdot \frac{T_0}{T}$, where $T_0 \cong 273.15\ K$ and $p_0 \cong 101,325\ kPa$. It follows from the above considerations that the breathing module $\varepsilon = \frac{\dot{v}_0}{\dot{V}_E}$ will be:

$$\varepsilon = 4 \cdot \frac{v_0 \cdot \dot{V}_{Me_2CO} \cdot d_{25°C}}{M_{Me_2CO} \cdot V_E \cdot \dot{n} \cdot \frac{p}{p_0} \cdot \frac{T_0}{T}} \tag{5.2}$$

where v_0 is molar volume of oxygen under normal conditions, $\dot{V}_{Me_2CO}$ is volume flow of *propanone*, $d_{25°C}$ is density of *propanone* at temperature $t = 25°C$, M_{Me_2CO} is molar mass of *propanone*, V_E is cylinder capacity of the piston breathing device, $\dot{n}$ is piston frequency of the respiratory device, p is pressure under measurement conditions, and t is temperature under measurement conditions.

As already mentioned in reaction (5.1), 1 *mol* of propanone Me_2CO reacts with 4 *mol* of oxygen O_2. Thus, one molar volume of acetone $V_{Me_2CO} \cong 73.89\ cm^3$ reacts with four molar volumes of oxygen $V = 4 \cdot v_0 \cong 97.76\ dm^3$. Thus, volume flow of consumed oxygen $\dot{v}$ can be calculated from proportion: $\dot{v} \cong \frac{97.76}{73.89} \cdot 10^3 \cdot \dot{V}_{Me_2CO} \cong 1323 \cdot \dot{V}_{Me_2CO}$. Inverse dependence can be written as $\dot{V}_{Me_2CO} \cong 0.756 \cdot 10^{-3} \cdot \dot{v}$. Hence, for *propanone* stream $\dot{V}_{Me_2CO}$ expressed in $\left[\dot{V}_{Me_2CO}\right] = cm^{-3} \cdot min^{-1}$ and flow of consumed oxygen $\dot{v}$ expressed in $\left[\dot{v}\right] = dm^3 \cdot min^{-1}$ the proportion takes the form: $\dot{v} \cong 1.323 \cdot \dot{V}_{Me_2CO}$. Assuming in compliance with Table 5.1 that the diver's oxygen consumption may be

TABLE 5.1
Example results of measurements of stable oxygen content x_s for various parameters of the respiratory and metabolic simulator settings

Depth	Temperature	Apparatus supply O_2/N_2	Parameters of respiratory simulator			Parameters of metabolic simulator		Stable content of oxygen	
			Volume of inspiration	Frequency	Respiratory module	Acetone volume	time	theoretical	measured
$[mH_2O]$	$[°C]$	$[m^3 \cdot m^{-3}]$	$[dm^3]$	$[min^{-1}]$	$[m^3 \cdot m^{-3}]$	$[cm^3]$	$[min : s]$	$[m^3 \cdot m^{-3}]$	$[m^3 \cdot m^{-3}]$
30	27.3	0.209	1.5	15	0.0702	21.0	17:14	0.072	0.072
40	28.3	0.209	1.5	15	0.0702	21.2	17:57	0.102	0.101
40	28.6	0.209	1.5	15	0.0703	21.2	17:57	0.102	0.101
50	28.5	0.209	1.5	15	0.0696	21.0	17:57	0.121	0.120
60	28.5	0.209	1.5	15	0.0682	21.4	18.40	0.135	0.137
30	29.4	0.209	1.5	15	0.0103	16.4	94:50	0.189	0.192
40	30	0.209	1.5	15	0.0105	10.2	58:10	0.193	0.195
50	31.5	0.209	1.5	15	0.0108	3.6	20:00	0.195	0.196
60	32.8	0.209	1.5	15	0.0117	5.0	25:50	0.196	0.198
40	31.1	0.209	1.5	15	0.0823	19.4	14:09	0.084	0.085
50	30	0.209	1.5	15	0.0826	20.4	14:46	0.104	0.102
50	29.6	0.209	1.5	15	0.0825	20.4	14.46	0.104	0.104
60	29.4	0.209	1.5	15	0.0800	19.8	14:46	0.122	0.120
30	19	0.209	2.5	15	0.0161	7.0	15:01	0.178	0.179

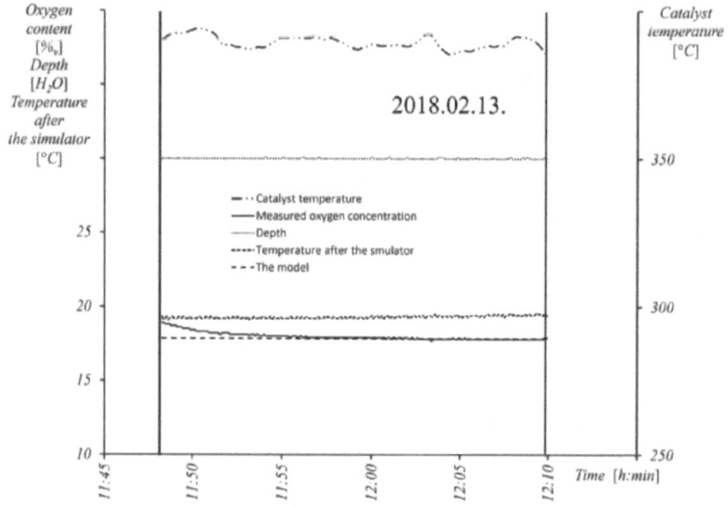

Parameters of the metabolic simulator				
Density of Me_2CO	ρ_{Me_2CO}	:	0.786 $g \cdot cm^{-3}$	25 °C
Molar mass of Me_2CO	M_{Me_2CO}	:	58.08 $g \cdot mol^{-1}$	
Dosing time of Me_2CO	t	:	15 min	1 s
Initial volume of Me_2CO	V_1	:	7 cm^{-3}	
Final volume of Me_2CO	V_2	:	0 cm^{-3}	

Parameters of the respiratory simulator			
Inspiratory / expiratory volume	V_E	:	2.5 dm^3
Number of breathing cycles	n	:	15 min^{-1}

Parameters of the diving apparatus			
Oxygen content in the premix	x_w	:	0.209 $mol \cdot mol^{-1}$
The ratio of bag volumes	r	:	0.104 $m^3 \cdot m^{-3}$

Parameters of the depth simulator			
Depth	H	:	**30 mH_2O**

The research results				
Molar volume of oxygen	V_{O_2}	:	23.95 $dm^3 \cdot min^{-1}$	19 °C
Oxygen consumption	$\dot{v}$	:	0.604 $dm^3 \cdot min^{-1}$	
Respiratory ventilation	$\dot{V}$	:	37.5 $dm^3 \cdot min^{-1}$	
Breathing module	ε	:	0.0161 $dm^3 \cdot min^{-1}$	
Theoretical stable oxygen content	x_{O_2}	:	0.178 $mol \cdot mol^{-1}$	
Measured oxygen content	x_{O_2}	:	0.179 $mol \cdot mol^{-1}$	

FIGURE 5.2 Example of results of investigations on ventilation in SCR CRABE SCUBA.

between $\dot{v} \in 0.3–2.4\,dm^3 \cdot min^{-1}$, the corresponding dosing of propanone Me_2CO should be within $\dot{V}_{Me_2CO} \in 0.23–1.81\,cm^3 \cdot min^{-1}$.

The current volumetric efficiency of the pump was measured during each measurement indirectly, by measuring the time with absolute error in time measurement $\Delta\tau = 0.2\ s$ used to empty the volume of the pipette with accuracy $\Delta V = \pm 0.2\ cm^3$.

Air and breathing mixtures were used to supply the SCR CRABE SCUBA. The results of the preliminary simulation tests are summarized in Table 5.1 together with the comparison of the theoretical values derived from the model (5.2). Figure 5.2 shows examples of the results from a single study.

SUMMARY

Simulation-based investigations are expected to allow determining the adequacy of the adopted model (4.3) of ventilation in the breathing space of the SCR with a metering bellows dispenser. Developing this model with the required accuracy provides the opportunity to work out data for the assumed decompression model.

As shown by the research data presented in Table 5.1, it can be expected that the model (4.3) is a sufficiently accurate approximation of the ventilation process of the diving apparatus and can be used to infer the parameters needed to plan the subsequent decompression.

NOTES

1 E.g., negative pressure of inspiration, positive pressure of expiration, breathing, changes in breathing resistance accompanying the change of the position of the apparatus in relation to the body of the diver, time of protective operation due to accumulated gas supply, etc.
2 E.g., weight in air and water, mutual distribution of mass and displacement elements and bending moments it produces, drag, etc.
3 E.g., availability of valves, ease of use and eliminating the possibility of errors, the possibility of maintenance and disinfection without tools, easy replacement of damaged components, ease of monitoring manometers and other measuring elements, visibility and the ability to communicate underwater through the mask, fitting the construction to the diver, validity of selecting materials, etc.
4 E.g., compliance with the documentation of the dimensions and masses of the components, the level of generated noise and magnetic field, the amount of dispersion of the gas medium released into water, resistance to corrosion and mechanical shock, time of protective action in relation to the ability to absorb carbon dioxide, etc.
5 E.g., harness system tensile strength, shock resistance of construction of flexible elements in the gas space to during dives into water, corrosion and fire resistance due to the increased oxygen content in the breathing gas, resistance of valves to frequent opening and closing, freezing resistance of breathing regulators, abrasion resistance of absorber fillings when shaken out, burst resistance of breathing bags, etc.
6 E.g., measuring elements, breathing regulators, absorber dosing nozzles, efficiency of ventilation of the breathing space, the possibility of accumulation of pollutants in breathing gas, etc.
7 Some designers use a special piston device to dispense carbon dioxide CO_2 in a ratio adequate to the amount of breathing gas taken, or simple dosing under the control of mass or rotametric dispensers, etc.

8 E.g., in investigations on ventilation in hyperbaric chambers.
9 E.g., hydrocarbon compounds.
10 ***Respiratory Quotient*** (*RQ*) during metabolizing of carbohydrates in the human body is determined at the level $\varepsilon_{RQ} = 1.0\ dm^3 \cdot dm^{-3}$, for proteins $\varepsilon_{RQ} \cong 0.8\ dm^3 \cdot dm^{-3}$, and for fat $\varepsilon_{RQ} \cong 0.7\ dm^3 \cdot dm^{-3}$.
11 Acetaldehyde.
12 Acetone.
13 Acetone.

REFERENCES

Frånberg O. 2015. Oxygen content in semi-closed rebreathing apparatuses for underwater use: Measurements and modeling. PhD diss., Stockholm School of Technology and Health. ISSN 1653-3836.

Kłos R. 2002. Metabolic simulator supports diving apparatus researches. *Sea Technology* 12, 53–56.

Kłos R. 2007. *Mathematical modelling of the normobaric and hyperbaric facilities ventilation.* Gdynia: Polish Hyperbaric Medicine and Technology Society. ISBN 978-83-924989-0-2.

Kłos R. 2009. *Wapno sodowane w zastosowaniach wojskowych.* Gdynia: Polish Hyperbaric Medicine and Technology Society. ISBN 978-83-92499889-5-7.

Kłos R. 2011. *Możliwości doboru dekompresji dla aparatu nurkowego typu CRABE.* Gdynia: Polish Hyperbaric Medicine and Technology Society. ISBN 978-83-924989-4-0.

Kłos R. 2012. *Możliwości doboru ekspozycji tlenowo-nitroksowych dla aparatu nurkowego typu AMPHORA – założenia do nurkowań standardowych i eksperymentalnych.* Gdynia: Polish Hyperbaric Medicine and Technology Society. ISBN 978-83-924989-8-8.

Kłos R. 2015. *Katalityczne utlenianie wodoru na okręcie podwodnym.* Gdynia: Polish Hyperbaric Medicine and Technology Society. ISBN 978-83-938-322-3-1.

Kłos R. 2016. *System trymiksowej dekompresji dla aparatu nurkowego typu CRABE.* Gdynia: Polish Hyperbaric Medicine and Technology Society. ISBN 978-83-938322-5-5.

Kyriazi N. 1986. *Development of an automated breathing and metabolic simulator.* Pittsburgh: Department of the Interior: USA Bureau of Mines, No. NC6414. 614.894.

Loncar M. 1992. Breathing simulator simulates human oxygen intake. *Offshore* 10, 82.

Part II

Hyperbaric Chambers

The second part of the monograph will describe the research on the processes of ventilation of hyperbaric complexes.

Hyperbaric chambers are much larger objects than diving apparatuses, hence the ventilation processes taking place in them run slower. However, deterministic models of the ventilation process of hyperbaric chambers are analogous to the models of ventilation of semi-closed-circuit breathing apparatus.

6 Ventilation of Hyperbaric Chambers

Ventilation of hyperbaric facilities[1] is performed during pressure exposures in order to:

- reduce carbon dioxide contents
- reduce oxygen leakage from internal assemblies of oxygen sets
- remove other contaminants from the atmosphere in hyperbaric facilities
- remove moisture
- reduce temperature[2] in the facility
- fulfill other tasks related to the type of diving.[3]

Ventilation control is performed to control carbon dioxide and contaminants, oxygen content, temperature, humidity,[4] and the stream of fresh breathing medium.[5] The problems discussed here concern diving on air, as well as oxygen used in the assemblies of oxygen sets.[6] The diver who uses a tight mouthpiece in an oxygen supply installation does not breathe air from the chamber and does not exhale oxygen into the chamber. As follows from the above-mentioned problem, during normal operation of oxygen sets BIBS[7] in a hyperbaric chamber, oxygen content should not increase.

A significant increase in temperature in a hyperbaric facility is caused by rapid pressurization of the facility. Intensive air ventilation of the container is one of the simple methods applied to reduce this temperature. In most cases, air temperature is lower than temperature of the gas inside the hyperbaric chamber. Additionally supplied fresh air from the pressure containers is expanded. According to *Joule-Thomson effect*, the temperature of the air supplied to the hyperbaric chamber decreases. Higher temperature air in the hyperbaric facility is partly flushed with the fresh, cooler breathing medium from the supplying tanks, and in the chamber the fresh gas is mixed with the hotter gas. The gas mixture then passes through the outlet valve. In order to decrease the temperature in the hyperbaric facility, the ventilation rate is usually maintained at the highest possible level. As temperature sensors are not expensive and widely available, the majority of chambers and hyperbaric complexes are fitted with them.

During decompression, water vapor usually occurs. In many situations, decreasing pressure in the hyperbaric facility causes cooling of the atmosphere.[8] At a dew-point, the water vapor occurs as fog. It causes discomfort to divers because the breathing medium has high humidity. It is very difficult for the servicing personnel to watch the divers because of deterioration in visibility caused by fog. In order to measure humidity in hyperbaric facilities, inexpensive hair hygrometers can be used.

Problems relating to contents of carbon dioxide, oxygen, and contaminants should be discussed separately. Carbon dioxide is hazardous because of its toxicity,[9] and oxygen is a fire hazard (Kłos, 2014). Oxygen at higher-pressure and long-duration exposures is toxic (Kłos, 2012). In order to monitor the contents of the above-mentioned gases, most often gas analyzers are used. By means of gas analyzers and other devices, the contents or partial pressure of oxygen, carbon dioxide, and contaminants are determined.

THE VENTILATION MODELS

Ventilation of a hyperbaric chamber differs from that of semi-closed-circuit rebreathers SCR. Mathematical models for a hyperbaric chamber ventilation can be derived using a similar procedure as in the case of SCR ventilation.

The investigations presented here were performed to check whether the mathematical models of ventilation derived for SCR can be also applied to hyperbaric chambers. The experimental results confirmed such a possibility.

The procedures proposed to employ to obtain mathematical models of ventilation can be used on both the micro and macro scales. The macro scale is concerned with the diving technique. It is important that the presented analytical models, compared to the empirical and semi-empirical models, are related to the physical phenomena occurring in the hyperbaric environment.

The schematic diagram of contaminant removal from hyperbaric facility atmosphere with the use of continuous ventilation is presented in the example of carbon dioxide[10] removal in Figure 6.1.

The streams of ventilating medium and oxygen consumed are usually related to NTP conditions, therefore the amount of carbon dioxide expressed in moles should

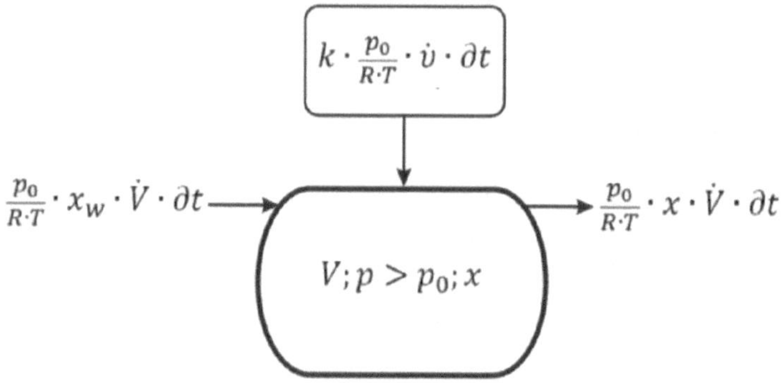

FIGURE 6.1 The molar balance of carbon dioxide in the habitat on the assumption that the carbon dioxide emission is equal to the oxygen consumption, where: k – the number of people in the hyperbaric chamber $\dot{V}$ – stream of the ventilating medium related to atmospheric pressure, $\dot{v}$ – stream of emitted carbon dioxide by the diver equal to oxygen consumption, x_w – carbon dioxide molar fraction in the fresh breathing medium, ∂t – elementary time, p – pressure corresponding to the depth, p_0 – normal pressure, R – universal gas constant, T – the absolute temperature, V – volume of the hyperbaric chamber.

be related to the expanded medium. It is assumed that the streams of ventilating medium, oxygen consumed, and breathing medium concerned with breathing action are not depth dependent (in fact, there is a small dependence on the depth [Kłos, 2000]). For continuous ventilation, differential and integral molar balance of carbon dioxide can be derived as presented in Tables 6.1–6.2.

TABLE 6.1
Derivation of the relation between carbon dioxide molar fraction *x* as a function of time *t* for the continuous ventilation of the habitat by means of integral calculus based on carbon dioxide molar balance in the chamber presented in Figure 6.1

Z:	1°	$\dot{V} \neq f(H);\ \dot{\upsilon} \neq f(H);\ \dot{V}_E \neq f(H)$	
	2°	$\dot{V} = \dot{V}_0 = \dot{V}(p = p_0);\ \dot{\upsilon} = \dot{\upsilon}_0 = \dot{\upsilon}(p = p_0)$	
T:		$x(O_2) \equiv x(CO_2) \equiv x = f(t)$	
P:	1°	$\frac{p}{R\cdot T}\cdot V\cdot\frac{\partial x}{\partial t} = \frac{p_0}{R\cdot T}\cdot\dot{V}\cdot x_w - k\cdot\frac{p_0}{R\cdot T}\cdot\dot{\upsilon} - \frac{p_0}{R\cdot T}\cdot\dot{V}\cdot x$	from the oxygen balance in the chamber in Figure 6.1
	2°	$\frac{\partial x}{\partial t} = \frac{p_0}{p}\cdot\frac{\dot{V}\cdot x_w}{V} + k\cdot -\frac{p_0}{p}\cdot\frac{\dot{v}}{V} - \frac{p_0}{p}\cdot\frac{\dot{V}}{V}\cdot x$ where: $a \equiv \frac{p_0}{p}\cdot\frac{\dot{V}\cdot x_w + k\cdot\dot{\upsilon}}{V} = idem; b \equiv \frac{p_0}{p}\cdot\frac{\dot{V}}{V} = idem$	as for the stable circumstances: $\dot{V}, \dot{v} = const$
	3°	$\forall_{a=\frac{p_0}{p}\cdot\frac{\dot{V}\cdot x_w - \dot{\upsilon}}{V}=idem;\, b=\frac{p_0}{p}\cdot\frac{\dot{V}-\dot{\upsilon}}{V}=idem}\ \partial x = (a - b\cdot x)\partial t$	from 2°
	4°	$\frac{\partial x}{b\cdot x - a} + \partial t = 0$	from 3° by dividing by $(b \cdot x - a)$
	5°	$\int\frac{\partial x}{b\cdot x - a} + \int\partial t = C$ where: C – integral constant	from 4° and integral definition
	6°	$\frac{1}{b}\cdot\ln\lvert b\cdot x - a\rvert + t = C; \ln\lvert b\cdot x - a\rvert \equiv b\cdot(C - t) = C' - b\cdot t$ where C' – the new constant	From 5°
	7°	$exp(C' - b\cdot t) = C''\cdot exp(-b\cdot t) \equiv b\cdot x - a$ where C′ – the new constant	from 6° and natural logarithm definition
	8°	If for boundary conditions $t \to 0 \Rightarrow x \to x_0$, then: $C'' = b\cdot x_0 - a \Rightarrow (b\cdot x_0 - a)\cdot exp(-b\cdot t) = b\cdot x - a$	From 7°
	9°	$x(t) = x_w + \frac{k\cdot\dot{\upsilon}}{\dot{V}} + \left(x_0 - x_w - \frac{k\cdot\dot{\upsilon}}{\dot{V}}\right)\cdot exp\left(-\frac{p_0}{p}\cdot\frac{\dot{V}}{V}\cdot t\right)$	from 2° i 8° q.e.d.

A – assumption, *T* – thesis, *P* – proof (evidence)
Where: k – the number of people in the hyperbaric chamber, $\dot{V}$ – stream of the ventilating medium related to atmospheric pressure, $\dot{v}$ – stream of emitted carbon dioxide by the diver equal to oxygen consumption, x_w – carbon dioxide molar fraction in the fresh breathing medium, ∂t – elementary time, x – molar fraction of carbon dioxide in the hyperbaric facility, p – pressure corresponding to the depth, p_0 – normal pressure, R – universal gas constant, T – the absolute temperature, V – volume of the hyperbaric chamber.

TABLE 6.2

Derivation of the relation between carbon dioxide molar fraction *x* as a function of time *t* for the continuous ventilation of the habitat by means of limits calculus based on carbon dioxide molar balance in the chamber presented in Figure 6.1

A:	1°	$\dot{V} \neq f(H);\ \dot{\upsilon} \neq f(H);\ \dot{V}_E \neq f(H)$	
	2°	$\dot{V} = \dot{V}_0 = \dot{V}(p = p_0);\ \dot{\upsilon} = \dot{\upsilon}_0 = \dot{\upsilon}(p = p_0)$	
T:		$x(O_2) \equiv x(CO_2) \equiv x = f(t)$	
P:	1°	For elementary time $d\tau$ at time$(i + 1) \cdot d\tau$ it can be written: $n(i+1) = (n(i) + \frac{p_0}{R \cdot T} \cdot \left[x_w \cdot \dot{V} + k \cdot \dot{\upsilon} - x(i) \cdot \dot{V}\right] \partial t$	from oxygen balance
	2°	$x(i+1) = \left[x(i) - x_w - \frac{k \cdot \dot{v}}{\dot{V}}\right] \cdot \left(1 - \frac{p_0}{p} \cdot \frac{\dot{V}}{V} \cdot \partial t\right) + x_w + \frac{k \cdot \dot{v}}{\dot{V}}$	from 1° divided by: $n = \frac{p \cdot V}{R \cdot T}$ and $x(i) = \frac{n(i)}{n}$
	3°	$x(i+2) = \left[x(i+1) - x_w - \frac{k \cdot \dot{v}}{\dot{V}}\right] \cdot \left(1 - \frac{p_0}{p} \cdot \frac{\dot{V}}{V} \cdot \partial t\right) + x_w + \frac{k \cdot \dot{v}}{\dot{V}}$	from 2°
	4°	$x(i+2) = \left[x(i+1) - x_w - \frac{k \cdot \dot{v}}{\dot{V}}\right] \cdot \left(1 - \frac{p_0}{p} \cdot \frac{\dot{V}}{V} \cdot \partial t\right)^2 + x_w + \frac{k \cdot \dot{v}}{\dot{V}}$	from 3° and 2°
	5°	$x(i+2) = \left[x(i+1) - x_w - \frac{k \cdot \dot{v}}{\dot{V}}\right] \cdot \left(1 - \frac{p_0}{p} \cdot \frac{\dot{V}}{V} \cdot \partial t\right)^j + x_w + \frac{k \cdot \dot{v}}{\dot{V}}$	form 2° ÷ 4°
	6°	For $i = 0$, $j = t$ and $x(0) = x_0$ it can be written: $x(t) = x_w + \frac{k \cdot \dot{v}}{\dot{V}} + \left(x_0 - x_w - \frac{k \cdot \dot{v}}{\dot{V}}\right) \cdot \lim_{j \to \infty}\left(1 - \frac{p_0}{p} \cdot \frac{\dot{V}}{V} \cdot \partial t\right)^j$	from 5°
	7°	$x(t) = a + b \cdot \lim_{j \to \infty}\left[\left(1 + \frac{c}{j}\right)^{j/c}\right]^c$ where: $a = x_w + \frac{k \cdot \dot{v}}{\dot{V}}, b = x_0 - a, c = -\frac{p_0}{p} \cdot \frac{\dot{V}}{V} \cdot \partial t$	from 6° $\lim_{j \to \infty}(j \cdot dt) = t$
	8°	$x(t) = exp\,(c) + a$, because: $\forall_{a(n) \neq 0} \lim_{n \to \infty} a(n) = 0 \Rightarrow \lim_{n \to \infty}\left(1 + a(n)\right]^{1/a(n)} \equiv e$	from 7° $a(n) = \frac{c}{j}$
	9°	$x(t) = x_w + \frac{k \cdot \dot{v}}{\dot{V}} + \left(x_0 - x_w - \frac{k \cdot \dot{v}}{\dot{V}}\right) \cdot exp\left(-\frac{p_0}{p} \cdot \frac{\dot{V}}{V} \cdot t\right)$	from 7°÷8° q.e.d.

A – assumption, *T* – thesis, *P* – proof (evidence)

Where: k – the number of people in the hyperbaric chamber, $\dot{V}$ – stream of the ventilating medium related to atmospheric pressure, $\dot{v}$ – stream of emitted carbon dioxide by the diver equal to oxygen consumption, x_w – carbon dioxide molar fraction in the fresh breathing medium, ∂t – elementary time, i – the successive time interval ∂t for substance balance, $x(i)$ – molar fraction of carbon dioxide in the hyperbaric facility at the moment $t = i \cdot \partial\tau$, p – pressure corresponding to the depth, p_0 – normal pressure, R – universal gas constant, T – the absolute temperature, V – volume of the hyperbaric chamber.

By this means the relationship between contaminant molar fraction and time is given:

$$x(t) = x_w + \frac{k \cdot \dot{\upsilon}}{\dot{V}} + \left(x_0 - x_w - \frac{k \cdot \dot{\upsilon}}{\dot{V}} \right) \cdot exp\left(-\frac{p_0}{p} \cdot \frac{\dot{V}}{V} \cdot t \right) \tag{6.1}$$

where $x(t)$ is carbon dioxide molar fraction at the moment t, t is time, and x_0 is carbon dioxide molar fraction at the moment $t = 0$.

An analysis of the Equation (6.1) gives:

$$x(t = 0) \equiv x_0 \tag{6.2}$$

$$x_s \equiv \lim_{t \to \infty} x(t) = x_w + \frac{k \cdot \dot{\upsilon}}{\dot{V}} \tag{6.3}$$

where x_s is the stable value of oxygen molar fraction in the chamber atmosphere $x_s = \lim_{t \to \infty} x(t)$.

The Equation (6.2) is a consequence of the assumed boundary conditions, whereas the Equation (6.3) follows from the mass balance.

MINIMUM AMOUNT OF GAS NECESSARY FOR CONTINUOUS VENTILATION

The derived equation describing the minimum amount of air necessary for continuous ventilation of the hyperbaric facility on the assumption that carbon dioxide partial pressure is maintained at a level not exceeding the permissible value (Przylipiak & Torbus, 1981; Kłos, 2014).

From the Equation (6.1) and the following assumptions[11]: $k = 1$ and $x_0 \equiv x_w$, it can be written: $x(t) = x_w + \frac{k \cdot \dot{v}}{\dot{V}} + \left(x_0 - x_w - \frac{k \cdot \dot{v}}{\dot{V}} \right) \cdot exp\left(-\frac{p_0}{p} \cdot \frac{\dot{V}}{V} \cdot t \right)$. When carbon dioxide partial pressure reaches the maximum permissible value p_{max} in time approaching infinity $t \to \infty$, then the value of ventilating medium stream reaches the minimum safe value $\dot{V}_{min}$ and from Equation (6.3) it can be written: $p_{max} = p \cdot x_w + p \cdot \frac{\dot{v}}{\dot{V}_{min}}$, where p represents the working pressure. Transformation of the last equation enables determining the minimum stream of ventilation medium in the course of continuous ventilation:

$$\dot{V}_{min} = \frac{p \cdot \dot{v}}{p_{max} - p \cdot x_w} \tag{6.4}$$

where $\dot{V}_{min}$ is the minimum stream of the ventilation medium, p_{max} is the maximum permissible carbon dioxide partial pressure, x_w is carbon dioxide molar fraction in the fresh breathing medium, $\dot{v}$ is the stream of carbon dioxide emitted by the diver, and p is working pressure.

CONTINUOUS VENTILATION

A. In accordance with the chemical branch standard for the compressed air for divers, the permissible carbon dioxide contents in the diving air is $0.05\%_v$, while for ideal gases it is equal to molar fraction having the value $0.0005\ mol \cdot mol^{-1}$.
B. In accordance with the operating manual of the hyperbaric chamber, type ORTOLAN L-80, the volume of the hyperbaric chamber is $V = 2.25\ m^3$. It is assumed that the chamber can be used for training up to the depth of $30\ mH_2O$.
C. On the assumption that the divers do not perform heavy work inside the decompression chamber, the stream of oxygen consumed is assumed to be approximately $\dot{v} = 1\,dm^3 \cdot min^{-1}$ (Kłos, 2012; Kłos, 2014). In such cases the respiratory quotient is to be assumed $0.8\ m^3 \cdot m^{-3}$. It is evident that when assuming that the oxygen stream consumed is at a level $\dot{v} = 1\,dm^3 \cdot min^{-1}$, the stream of carbon dioxide emitted by the "standard man" (light work) is approximately $\dot{v} = 0.8\,dm^3 \cdot min^{-1}$. Assuming that during medical treatment and decompression divers perform light work equal to $\dot{v} = 0.5\,dm^3 \cdot min^{-1}$, the theoretical average carbon dioxide emission will be at a level of $\dot{v} = 0.4\,dm^3 \cdot min^{-1}$.
D. Pursuant to the navy diving regulations, the maximum permissible carbon dioxide partial pressure for the people in a hyperbaric facility during decompression is equal to $1.5\ kPa$.

Summarizing the above, assessments and Equation (6.4) conclusions that can be drawn are presented in Table 6.3 and Figure 6.2.

INTERRUPTED VENTILATION

Continuous ventilation is not generally accepted in the diving methodology except ventilation of the breathing space in the diving helmet used by the classic diver, as well as ventilation of the habitat in the closed system.[12] Interrupted ventilation is most often used in the open ventilation system in hyperbaric facilities. The idea of

TABLE 6.3
The table of continuous ventilation of the hyperbaric diving base type ORTOLAN L-80

The depth H	Approximate pressure at the given depth p	The minimum stream of ventilating medium $\dot{V}_{min}$	
		1 person	2 persons
$[mH_2O]$	$[kPa]$	$[dm^3 \cdot min^{-1}]$	
10	200	57.1	114.3
20	300	88.9	117.8
30	400	123.1	246.1

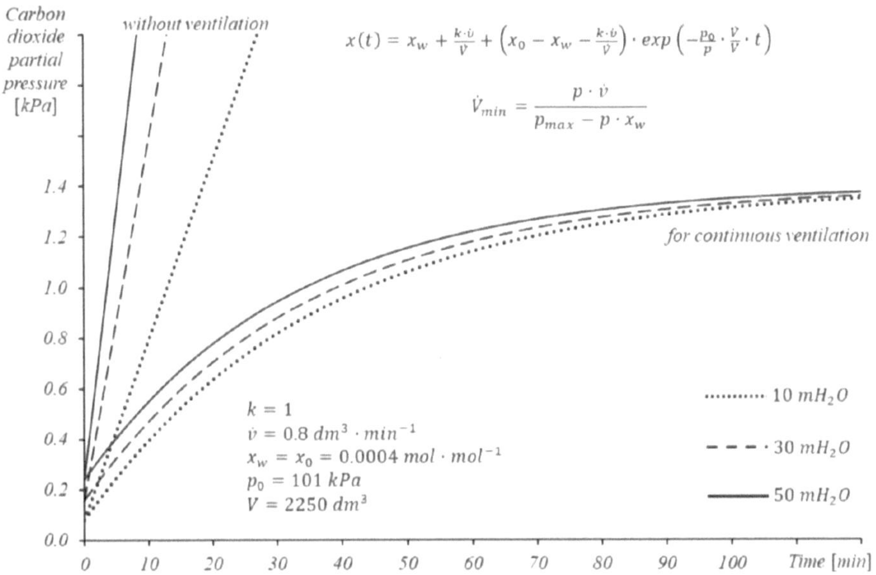

FIGURE 6.2 The diagram of the relationship between carbon dioxide partial pressure and time for continuous ventilation of the chamber ORTOLAN L-80 shows the minimum calculated stream of ventilating air having a carbon dioxide content of $0.04\%_v CO_2$ was set for different total pressures; where: k – the number of people in the hyperbaric chamber, $\dot{V}$ – stream of the ventilating medium related to atmospheric pressure, $\dot{V}_{min}$ – minimum stream of the ventilation medium, $\dot{v}$ – stream of emitted carbon dioxide by the diver equal to oxygen consumption, p – pressure corresponding to the depth, p_0 – normal pressure, V – volume of the hyperbaric chamber, p_{max} – maximum permissible carbon dioxide partial pressure, x_w – carbon dioxide molar fraction in the fresh breathing medium, $x(t)$ – carbon dioxide molar fraction at the moment t, x_0 – carbon dioxide molar fraction at the moment $t = 0$, t – time

continuous ventilation[13] was dropped because of technical difficulties[14] and discomfort[15] it causes. The ventilating medium should be contaminant free or its content should be low as compared with the permissible contaminant content. For this reasons the breathing medium used for ventilation should be checked and should be certified. In the course of pressurizing a diver to the depth of plateau with the breathing mixture having the properties that meet the above requirements, there is some time during which the chamber may be not ventilated. The time from the beginning of the decompression of a diver to the beginning of ventilation is known as "time to the first ventilation."

As before, the calculations will be based on the example of carbon dioxide removal. Carbon dioxide molar balance for the case of an unventilated habitat is presented in Figure 6.3.

It enables derivation of the relationship between carbon dioxide molar fraction and time and then to find time to the first ventilation t_{max}:

$$t_{max} = \frac{V}{k \cdot \dot{v}} \cdot \frac{p_{max} - p \cdot x_0}{p_0} \tag{6.5}$$

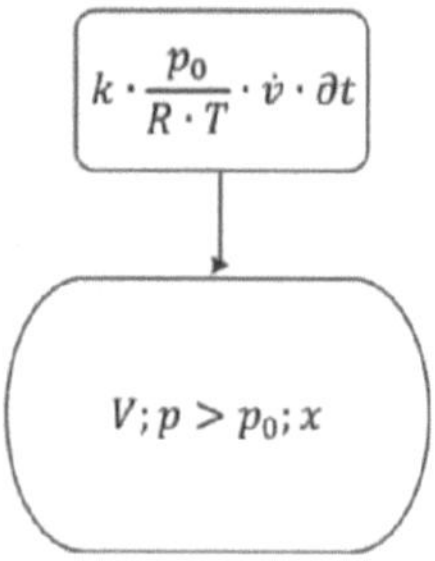

FIGURE 6.3 Carbon dioxide molar balance for the unventilated habitat.

where k is the number of people in the hyperbaric chamber, t_{max} is time to the first ventilation, p_{max} is maximum permissible carbon dioxide partial pressure, p_0 is normal pressure, $\dot{v}$ is the stream of carbon dioxide emitted by the diver, p is working pressure, and V is volume of the hyperbaric chamber.

Ventilation intensity can be determined from the Equation (6.1). Emission of contaminant is neglected $\dot{v} \cong 0$ due to the high rate of ventilation. Hence, the amount of air that is needed for interrupted ventilation V_0 can be written as:

$$\forall_{k \cdot \dot{v} \ll \dot{V}} \; V_0 = \dot{V} \cdot t = \frac{p_0}{p} \cdot V \cdot ln\left(\frac{x_0 - x_w}{x - x_w}\right) \tag{6.6}$$

where V_0 is the amount of ventilating air needed at the *NTP* conditions, t is time of ventilation, and x is the final contaminant molar fraction in the chamber after the interrupted ventilation has been finished.

It is necessary to ventilate the chamber in the shortest possible time.

Time to the next ventilation is determined from Equation (6.3), where x_0 equals to carbon dioxide contents after the previous ventilation has been completed. In order to observe determination of the ventilation table, the following example will be given (Table 6.4):

A) In accordance with the chemical branch standard for the compressed air for divers, the permissible carbon dioxide contents in the diving air is $0.05\%_v$, and for ideal gases it is equal molar fraction of the value 0.0005 $mol \cdot mol^{-1}$.
B) In accordance with the operating manual of the hyperbaric chamber, type ORTOLAN L-80, volume of the hyperbaric chamber is $V = 2.25\ m^3$. It is assumed that the chamber can be used for training up to the depth of 30 mH_2O.
C) On the assumption that the divers do not perform heavy work inside the decompression chamber, the stream of oxygen consumed is assumed to be approximately $\dot{v} = 1\,dm^3 \cdot min^{-1}$ (Kłos, 2012; Kłos, 2014). In such cases the respiratory quotient is to be assumed as 0.8. It is evident that when assuming that the oxygen stream consumed is at a level $\dot{v} = 1\,dm^3 \cdot min^{-1}$, the stream of carbon dioxide emitted by the "standard man" (light work) is approximately equal to $\dot{v} = 0.8\,dm^3 \cdot min^{-1}$. Assuming that during medical treatment and decompression the divers perform light work of $\dot{v} = 0.5\,dm^3 \cdot min^{-1}$, the theoretical average carbon dioxide emission will be at a level of $\dot{v} = 0.4\,dm^3 \cdot min^{-1}$.

TABLE 6.4
The calculation results of time to the first ventilation t_{max}, ventilation air demand V_0, and time for the next ventilation t

The depth	The pressure	Time to the first ventilation	Ventilation air demand	Time to the next ventilation
H	p	t_{max}	V_0	t
$[mH_2O]$	$[kPa]$	$[min]$	$[m^3]$	$[min]$
10	200	39.4	1.988	14.1
20	300	38.0	3.123	14.1
30	400	36.6	4.370	14.1
40	500	35.2	5.747	14.1
50	600	33.8	7.276	14.1
60	700	32.3	8.986	14.1

D) In accordance with the Polish Navy diving regulations, the maximum permissible carbon dioxide partial pressure for the people in a hyperbaric facility during decompression is equal to 1.5 kPa.
E) It is assumed that once the ventilation has been completed, the carbon dioxide partial pressure will decrease to the level of 1 kPa.

Time to the first and further ventilation t_{max} is determined from Equation (6.5) and ventilation intensity $\dot{V}$ and air demand V_0 can be calculated from Equation (6.6). The results of calculations are given in Table 6.4.

GAS MIXTURE METERING

It was decided to use the gas constant metering set as the carbon dioxide simulator. It consists of a pressure reducer and a gas mixture metering nozzle. The mass stream of gas is effected by the gas composition, the backward pressure, and the value of reduced pressure.

It was experimentally proved that the stream of flowing gas medium was constant when the proportion of the pressure between the inlet and outlet of the nozzle exceeded a certain value, despite an increase in pressure on the outlet up to the limiting value: the backward pressure. In other words, an increase in the backward pressure below a certain limit will not decrease the stream of the flowing medium. It is an operating principle of metering nozzles. In order to deliver the constant stream of gas to the assumed maximum backward pressure, it is necessary to choose an appropriate pressure reducer set and a metering nozzle. In order to find the necessary equations, *the Saint-Venanta-Wantzela equation* was derived from the *Bernouilli equation* (Goliński & Troskalański, 1979; Mittleman, 1989; Kłos, 2015):

$$\dot{V} = \Psi \cdot A \cdot \sqrt{\frac{R \cdot T}{M}} = \Psi \cdot A_0 \cdot \sqrt{\frac{p_1}{\rho_1}} \quad \Psi = \sqrt{2 \cdot \frac{\kappa}{\kappa - 1} \cdot \left[\left(\frac{p_2}{p_1} \right)^{2/\kappa} - \left(\frac{p_2}{p_1} \right)^{\kappa+1/\kappa} \right]} \tag{6.7}$$

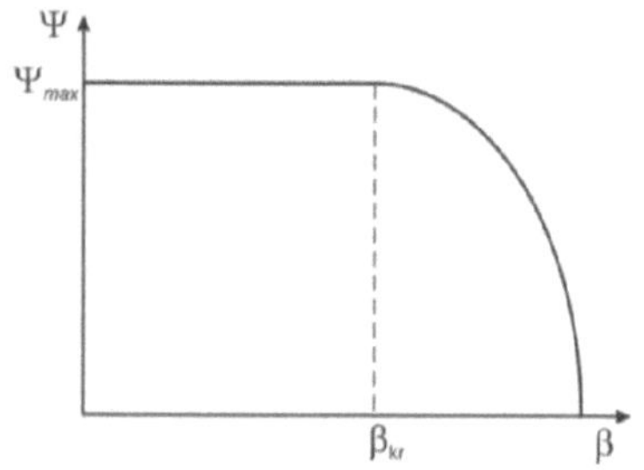

FIGURE 6.4 The practical representation of the flow number Ψ *vs* the ratio $\beta = \frac{p_2}{p_1}$, where p_2 is the pressure after nozzle and p_1 represents the pressure before the nozzle.

where $\dot{V}$ is the volume stream of the gas metering related to conditions during measurements before the nozzle, A is cross-section area of the nozzle, R is universal gas constant, T is thermodynamic temperature, M is average molar mass of gas mixture, Ψ is the flow number, κ is adiabatic exponent, p_1 is the pressure before the nozzle, p_2 is the pressure after the nozzle.

The graphic representation of the flow number Ψ in relation to the ratio $\beta = \frac{p_2}{p_1}$ of the pressure after p_2 to the pressure before p_1 is a straight line of the maximum at the critical value β_{kr} (Figure 6.4).

The value β_{kr} is determined from the derivative $\frac{\partial \Psi}{\partial \beta} \equiv 0$ that equals zero. Hence, it can be written: $\beta_{kr} = \left(\frac{2}{\kappa+1}\right)^{\kappa/\kappa-1}$. The value Ψ_{max} can be determined from the following equation:

$$\Psi_{max} = \sqrt{\kappa \cdot \left(\frac{2}{\kappa+1}\right)^{\kappa+1/\kappa-1}} \tag{6.8}$$

where Ψ_{max} is maximum value of flow number Ψ.

The values of adiabatic exponent κ, critical pressure ratio β_{kr}, critical flow number Ψ_{max}, and molar masses M for typical components of the breathing media are presented in Table 6.5.

Using the value of pressure ratio β_{kr} from Table 6.5, the minimum nozzle supplying pressure p_1 to atmospheric pressure p_0 with the use of air: $p_1 = \frac{p_0}{\beta_{kr}} = \frac{100\,kPa}{0.528} \cong 0.2\,MPa$ can be calculated. Relating the gas stream to *NTP conditions* $\dot{V}_0 = \frac{p_b}{p_0} \cdot \frac{T_0}{T} \cdot \dot{V}$ can be written, where $\dot{V}$ represents the measured value of the metered gas stream, p_b the barometric pressure, p_0 the normal pressure, T the thermodynamic temperature during experiments, and T_0 the normal temperature. The slope of the straight line that represents the function $\dot{V}_0 = f(p_1)$ equals the derivative of the stream with respect to the supplying pressure:

TABLE 6.5
The values of the chosen parameters for the components of the typical breathing gases

	κ	β_{kr}	Ψ_{max}	M [$kg \cdot mol^{-1}$]
air	1.4	0.528	0.685	28.96
oxygen	1.4	0.528	0.685	31.999
nitrogen	1.4	0.528	0.685	28.01
helium	1.677	0.486	0.728	4.003
CO_2	1.3	0.546	0.667	44.01

$$\left(\frac{\partial \dot{V}_0}{\partial p_1}\right)_{\beta<\beta_{kr}} = \frac{T_0}{T} \cdot \frac{\Psi_{max}}{p_0} \cdot \sqrt{\frac{R \cdot T}{M}} \cdot A \tag{6.9}$$

hence:

$$A = \frac{\pi \cdot D^2}{4} = \frac{p_0 \cdot T}{T_0 \cdot \Psi_{max} \cdot \sqrt{\frac{R \cdot T}{M}}} \left(\frac{\partial \dot{V}_0}{\partial p_1}\right)_{\beta<\beta_{kr}} \tag{6.10}$$

It was assumed that the measurement temperature was $20°C$. Other measurement temperatures introduce a slight error in the calculated value. The fixed temperature makes it possible to design a chart used to construct a nozzle. Therefore, the Equation (6.10) can be a base for designing of the nozzles for the reduce-metering sets of a semi-closed-circuit diving apparatus with the constant gas metering. For this purposes, the following function can be written:

$$\forall_{\beta<\beta_{kr}} \left(\frac{\partial \dot{V}_0}{\partial p_1}\right)_Y = f(D) \tag{6.11}$$

where D is the nozzle diameter and Y is the kind of the gas mixture.

An advantage of the presented method of calculation is that it enables simple recalculation of gas mixture metering Y. As it follows from the above, in order to find the metering rate of gas mixtures, it is necessary to measure accurately metering of the air nozzle. Therefore, the preliminary process of the measurements is simplified. The results are given in Table 6.6.

Transforming the Equation (6.11) and introducing the unit conversion factors, and taking advantage of the fact that the beginning of the function diagram passes through the point (0, 0), the following can be written:

$$\forall_{\beta<\beta_{kr}} \left(\frac{\dot{V}_0}{p_1}\right)_Y \cong 685 \cdot \frac{\Psi_{max}}{\sqrt{M}} \cdot D^2 \tag{6.12}$$

TABLE 6.6
The results obtained from Equation (6.12) for pure gases and gas mixtures at the measuring temperature 20°C; the gas stream is related to NTP conditions

	$\left(\frac{\partial \dot{V}_0}{\partial p_1}\right)_{20°C} \left[\frac{dm^3}{MPa \cdot min}\right]$						
$D[mm] \rightarrow$	**0.1**	**0.2**	**0.3**	**0.4**	**0.5**	**0.6**	**Formula** $\left(\frac{\partial \dot{V}_0}{\partial p_1}\right)_{20°C}$
Air	0.86	3.44	7.75	13.77	21.53	31.00	$86.11\ D^2$
Helium	2.46	9.85	22.15	39.39	61.54	88.62	$246.16\ D^2$
$50\%_v$ air in helium	1.18	4.71	10.60	18.85	29.45	42.41	$117.82\ D^2$

where $[M] = kg \cdot kmol^{-1}$, $[p_1] = MPa$, $\left[\dot{V}_0\right] = dm^3 \cdot min^{-1}$, and $[D] = mm$.

Measurements of the fresh breathing medium metering with the use of the designed nozzle will be omitted because they have been described previously (Kłos, 2007a; Kłos, 2007b).

THE DESIGN OF A CARBON DIOXIDE EMISSION SIMULATOR

In order to perform the investigations on the hyperbaric chambers, it was necessary to build a carbon dioxide emission simulator. The design and engineering realization of the simulator were made with the purpose of furthering preliminary investigations. These preliminary investigations had to be carried out in order to determine the possibilities to use the simulator as a simulator of carbon dioxide exhaled by divers in the hyperbaric environment. An external disposer, type MVB, that supplies the SCR FGG III UBA was used as the basis of the construction[16] (Figure 6.5).

Other devices of this type can be adopted, for example, an external disposer of an experimental version of the diving apparatus type SCR GAN-87, or the disposers of the other types of SCR with the constant gas metering. It is important to make sure that the metering nozzle is easy to exchange or switch by the proper valve.

It was assumed for the preliminary design that carbon dioxide emission should be maintained at the level of 0.8 $dm^3 \cdot min^{-1}$ for NTP conditions within the depth range up to 60 mH_2O. Cooling and solidification of pure carbon dioxide may occur due to its flow through the small nozzles.

Therefore, a decision was made to use the gas mixture $50\%_v CO_2/air$. It is recommended that the mixture pressure be maintained at a level not exceeding 7 *MPa* because at higher pressures carbon dioxide exists in the liquid state. In order to avoid stratification, the gas mixture should be well homogenized and stored in a horizontal position. Before use, the gas mixture should be repeatedly mixed by rotating, heating, and cooling of the gas cylinder. The calculated values of the necessary parameters derived from the adopted assumptions and the calculation results[17] are described in Table 6.7.

The nominal supplying pressure is equal to approximately 2.0 *MPa*, and nitrogen metering is maintained at the level of 1.8 $dm^3 \cdot min^{-1}$ These values were determined

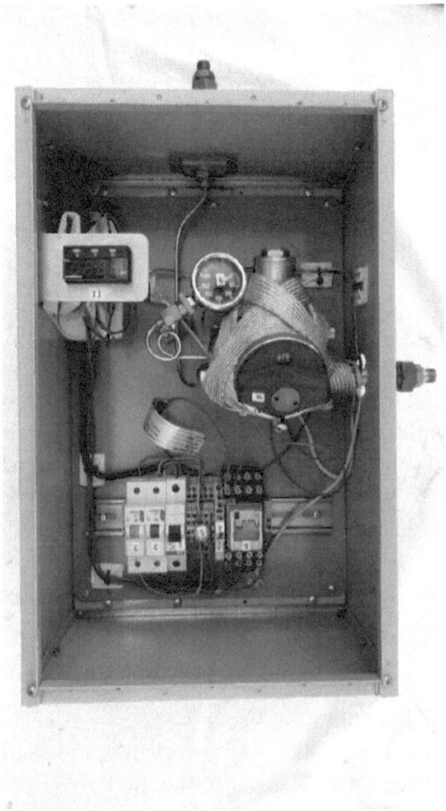

FIGURE 6.5 An external disposer of the type MVB adopted as the carbon dioxide emission simulator.

TABLE 6.7
The preliminary calculations for the carbon dioxide emission simulator

The gas mixture composition			The values of the parameters						The streams of gases		
O_2	N_2	CO_2	M	κ	ρ	Ψ_{max}	β_{kr}	p_1	A	B	C
$[\%_v]$			$[kg \cdot kmol^{-1}]$	–	$[kg \cdot m^3]$	–	–	[MPa]	$[dm^3 \cdot min^{-1}]$		
10.5	**39.5**	**50.0**	**36.429**	**1.350**	**1.633**	**0.676**	**0.537**	**1.4**	**1.6**	**1.8**	**1.8**

where:
- A The real stream of gas mixture $50\%_v CO_2/air$
- B Indications of rotameter that has been scaled with the respect to air at planned nominal flow of the gas mixture $50\%_v CO_2/air$
- C An equivalent nitrogen stream flowing through the nozzle of the nominal stream of the gas mixture $50\%_v CO_2/air$

from the preliminary calculations of the simulator design. The carbon dioxide simulator built was tested in the laboratory settings. The results are presented in Table 6.8.

Consequently, gas mixture $50\%_v CO_2/air$ was used in the research carried out. The results of the experiments are presented in Table 6.9.

As it follows from the comparison of the values assumed in the design and experimental results, it is possible to design any simulator of carbon dioxide emission. In order to perform research on hyperbaric chamber ventilation, a carbon dioxide simulator should be designed.

INVESTIGATIONS ON HYPERBARIC CHAMBER VENTILATION

The decompression chamber ORTOLAN L-80 was an experimental object (Figure 6.6).

It was used as an object for necessary calculations and pilot experiments. According to the servicing principles of this type diving base, the hyperbaric chamber has the volume of 2.25 m^3 and can be used as a training chamber up to the depth 30 mH_2O.

The chamber is fitted with two outlet valves, for rapid and controlled emptying. The valve of the controlled dump is connected with the rotameter having the range 50–400 ±20 $dm^3 \cdot min^{-1}$ @ $t = 0°C \wedge p = 101.325$ kPa. It can be used to control continuous ventilation. Due to high intensity of interrupted ventilation, it is necessary to use the valve of rapid emptying. For its suitability it is necessary to find the flow characteristic of the valve.

The rapid emptying valve used in the chamber, type L-80, is not connected to any device that controls the medium flow. For this reason it is necessary to determine its flow characteristics. Based on the flow characteristics, it is possible to find the stream of the flowing breathing medium. The simple method used to calculate flow capacity is presented in (US Navy Diving Manual, 1980; US Navy Diving Manual, 2016). The stream of gas is determined by measuring pressure drop time at a specific open angle[18] of the outlet valve. The following procedure was adopted to determine the flow capacity.[19] The outlet valve was marked.[20] This enabled opening it by making a fixed number of turns or part of a turn with one-fifth of turn accuracy. These measurements were omitted because they were described in detail earlier (Kłos, 2000).

The example of the measurements is presented in Table 6.10.

The recorded differences in temperature are not significant as the temperature change by 3°C produces approximately 1% change in the gas stream.

In order to calculate the stream of the gas flowing through the open valve, a mass balance of the hyperbaric chamber was assumed (Figure 6.1 and Table 6.2). Decrease in the amount of gas moles in the hyperbaric chamber at the pressure drop is determined from the *Clapeyron equation*:

$$\Delta n = n_p - n_k = \frac{V}{R \cdot T} \cdot \Delta p = \frac{V}{R \cdot T} \cdot \Delta p \cdot \left(p_p - p_k \right) \tag{6.13}$$

where Δn is decrease in gas amount in the hyperbaric chamber, n_p is the initial number of the gas moles in the hyperbaric chamber, n_k is the final number of the gas moles in the hyperbaric chamber, V is volume of the chamber, p_p is the initial pressure, p_k is the final pressure, and Δp is pressure drop.

TABLE 6.8
The results of the testing measurements of nitrogen emission from the designed and built nozzle of the carbon dioxide simulator

Measurement conditions		Barometric pressure $p_a = 102.4\ kPa$		The temperature $t = 24.0°C$ The absolute temperature $T = 273.15 + t = 297,15K$			The kind of the breathing mixture: nitrogen Density of the breathing medium: $\rho = 1.251\ kg \cdot m^{-3}$			
The depth	The supplying pressure	The reduced pressure	The rotameter indications	Metering (rotameter)	Dosing (rotameter)	Volume of the bell	The bell filling up time	Dosing (the bell)	Dosing (the bell)	Remarks
H	p	p_z	$\dot{V}$	$\dot{V}_0$	-	V_d	t	$\dot{V}_0$	-	
$[mH_2O]$	$[MPa]$	$[MPa]$	$[dm^3 \cdot h^{-1}]$	$[dm^3 \cdot min^{-1}]$	$[dm^3 \cdot min^{-1}]$	$[dm^3]$	$[min : s]$	$[dm^3 \cdot min^{-1}]$		
±1	±0.8	±0.06	±20	±0.4	-	±0.5	±0.2"		±0.6	
0	12.3	0.21	110	1.81		10	5:2.0	1.84		
10	12.2	0.21	110	1.81	$\dot{V}_0 \cong [1.8 \pm 0.4]\ dm^3 \cdot min^{-1}$				$\dot{V}_0 \cong [1.8 \pm 0.6]\ dm^3 \cdot min^{-1}$	nr: MVB 145
20	12.2	0.21	109	1.80						
30	12.1	0.21	106	1.75						Rotameter MLW of the range 200–2000 $\pm 20\ dm^3 \cdot h^{-1}$
40	12.1	0.21	106	1.75						
50	12.0	0.21	104	1.71						
60	12.0	0.21	102	1.68						
70	12.0	0.21	102	1.68						
0	12.0	0.21	109	1.80						

TABLE 6.9

The test measurements of gas mixture 59%$_v$$CO_2$/*air* emission; the other devices of this type can be adopted the designed and built of the carbon dioxide simulator nozzle

Measurement conditions		Barometric pressure $p_a = 102.4\ kPa$		The temperature $t = 24.0°C$ The absolute temperature $T = 273.15 + t = 297{,}15K$			The kind of the breathing mixture: nitrogen Density of the breathing medium: $\rho = 1.251\ kg \cdot m^{-3}$			
The depth	The supplying pressure	The reduced pressure	The rotameter indications	Metering (rotameter)	Dosing (rotameter)	Volume of the bell	The bell filling up time	Dosing (the bell)	Dosing (the bell)	Remarks
H	p	p_z	$\dot{V}$	$\dot{V}_0$	-	V_d	t	$\dot{V}_0$	-	
$[mH_2O]$	$[MPa]$	$[MPa]$	$[dm^3 \cdot h^{-1}]$	$[dm^3 \cdot min^{-1}]$	$[dm^3 \cdot min^{-1}]$	$[dm^3]$	$[min : s]$	$[dm^3 \cdot min^{-1}]$		
±1	±0.8	±0.06	±20	±0.4	-	±0.5	±0.2"		±0.6	
0	7.8	0.21	112	1.62		10	5:2.0	1.71		
10	7.8	0.21	111	1.60	$\dot{V}_0 \cong [1.6 \pm 0.4]\ dm^3 \cdot min^{-1}$	10	5:2.8	1.71	$\dot{V}_0 \cong [1.7 \pm 0.6]\ dm^3 \cdot min^{-1}$	nr: MVB 145
20	7.8	0.21	110	1.59						
30	7.7	0.21	110	1.59						
40	7.7	0.21	109	1.57						Rotameter MLW of the range 200–2000 ±20] $dm^3 \cdot h^{-1}$
50	7.7	0.21	108	1.56						
60	7.7	0.21	107	1.54						
70	7.6	0.21	106	1.53						
0	7.6	0.21	110	1.59						

From Equation (6.13) it follows that the molar stream of gas can be calculated as:

$$\dot{n} = \frac{\Delta n}{\Delta t} = \frac{V}{R \cdot T} \cdot \frac{\Delta p}{\Delta t} \tag{6.14}$$

where $\dot{n}$ is the stream of moles and t is pressure drop time.

Based on *Clapeyron equation*, the following can be written:

$$\dot{V} = \dot{n} \frac{R \cdot T}{p_0} \tag{6.15}$$

FIGURE 6.6 Mobile diving system ORTOLAN L-80

(*Continued*)

(c)

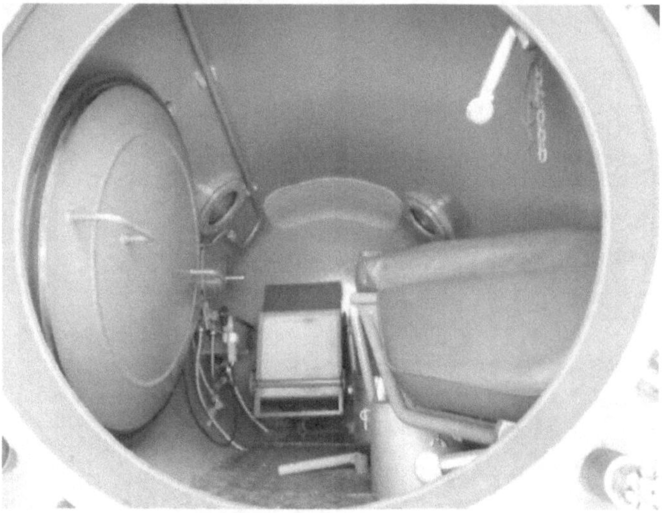

(d)

FIGURE 6.6 (Continued) Mobile diving system ORTOLAN L-80.

where $\dot{V}$ is the ventilation stream related to barometric pressure p_0 and p_0 is barometric pressure.

It follows from Equations (6.14) and (6.15) that:

$$\dot{V} = \frac{V}{p_0} \cdot \frac{\Delta p}{\Delta t} \tag{6.16}$$

TABLE 6.10
The calculation results of the gas capacity flow through the rapid emptying valve of the chamber

The overlap angle of the valve		Pressure drop time by $\Delta p = 3\ mH_2O$ for chosen depth [s]								
		$10\ mH_2O$			$20\ mH_2O$			$30\ mH_2O$		
The number of the valve turns	The angle	Average Pressure drop time	Tempera-ture	The ventilation steam	Average Pressure drop time	Tempera-ture	The ventilation steam	Average Pressure drop time	Tempera-ture	The ventilation steam
	[°]	[s]	[$°C$]	[$m^3 \cdot min^{-1}$]	[s]	[$°C$]	[$m^3 \cdot min^{-1}$]	[s]	[$°C$]	[$m^3 \cdot min^{-1}$]
0.4	144	52.1	24.4	0.78	29.1	24.2	1.39	17.2	24.0	2.35
0.8	288	43.1	24.4	0.94	25.9	24.2	1.56	16.3	24.0	2.48
1.2	432	39.7	24.2	1.02	23.2	24.0	1.75	14.4	23.8	2.81
1.6	576	37.6	24.4	1.08	22.8	24.2	1.78	13.5	23.8	3.00
2.0	720	37.5	24.6	1.08	22.6	24.2	1.79	13.4	23.8	3.02
2.4	864	36.8	24.6	1.10	22.9	24.2	1.77	13.1	24.0	3.09
2.8	1008	37.3	24.6	1.09	22.9	24.4	1.77	12.8	24.0	3.18
3.2	1152	36.4	25.0	1.11	22.7	24.8	1.78	12.8	24.6	3.16
3.6	1296	37.6	25.4	1.08	22.4	25.2	1.81	13.3	24.6	3.06

The chamber volume $V = 2.25\ m^3$

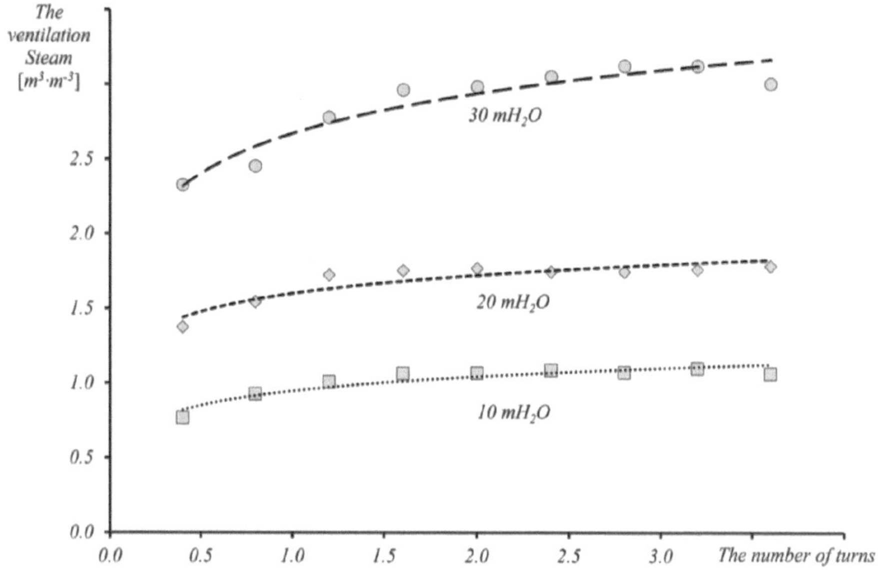

FIGURE 6.7 The flow characteristics of the rapid emptying valve of the ORTOLAN L-80 chamber.

Based on Equation (6.16), the flow capacities of the gas were determined for the specified valve set-points. The results are presented in Table 6.10 and in a graphic form in Figure 6.7.

The comparison of the results obtained in the course of determining the flow capacity of the rapid emptying valve of the chamber ORTOLAN L-80 with the demand air ventilation table – Table 6.4 can be used to draw relevant conclusions.

Interrupted ventilation should be performed in the shortest possible time. It follows from the comparison of the results that the valve should be opened for the greatest flow capacity independently of the depth of diving. The greatest flow capacity is depth independent[21] and occurs at the valve set-point that corresponds to 2¼ turns of the valve hand-wheel.

The ventilation stream, minimum air requirements, and time for interrupted ventilation are presented in Table 6.11. As it follows from the theoretical calculations, the minimum time of the interrupted ventilation of the chamber, type ORTOLAN L-80, is approximately 2 *min*.

THE PRELIMINARY INVESTIGATIONS ON INTERRUPTED VENTILATION

In order to verify the correctness of the assumed model of interrupted ventilation simulation, diving with participation of divers is needed. Two divers were compressed to the pressure corresponding to the diving depth[22] of 30 mH_2O. To maintain the proper homogenization of the ORTOLAN L-80 chamber atmosphere, the divers mix it using towels held in their hands.

TABLE 6.11
The theoretical data of interrupted ventilation for the *ORTOLAN L* – 80 chamber

Depth	Time to the first ventilation	Ventilation air requirement	The ventilation stream at the rapid emptying valve†	Minimum ventilation time	Maximum ventilating time
$[mH_2O]$	$[min]$	$[m^3]$	$[m^3 \cdot min^{-1}]$	$[min: s]$	$[min: s]$
10	39.4	2.0	1.1	1:49	14:06
20	38.0	3.1	1.8	1:44	14:06
30	36.6	4.4	3.1	1:25	14:06

† Set-point corresponding to a 2¼ turn
Calculation of air ventilation requirements and maximum time to the next ventilation was performed according to the assumptions presented in Table 6.4

Gas was continuously sampled from the chamber and analyzed with regard to carbon dioxide. The gas analyzer, type Analox 500, was used. The metrological properties of the gas analyzers are presented in Table 6.12.

A short and thin gas sampling line was applied. Therefore, the delay time up to 20 *s* was not significant. The amount of sampled gas compared with the chamber volume was not significant. It was approximately 0.04%. Except for the measurements of the carbon dioxide CO_2 contents, the following parameters had to be determined:

TABLE 6.12
The gas analyzer, type ANALOX 5000, properties

Specification	The value
Measuring range:	0–1$\%_v CO_2$
Resolution:	0.001$\%_v CO_2$
The measurement accuracy:	0.01$\%_v CO_2$ (1% *F. S.*)
Linearity:	1% *F. S.*
Repeatability:	1% *F. S.*
The typical response time:	3 *s*
The typical zero stability:	1% *F. S.* per week
The temperature of the optical bench:	60°*C*
Heating time of the optical bench:	up to 15 *min* at the temperature 20°*C*
The gas flow range:	50÷200 $cm^3 \cdot min^{-1}$
Pressure error effect:	0.15% of the measured value per1 mbar
Pressure of the analysed gas sample:	±350 mbar overpressure/under pressure
The temperature range:	5÷50°*C*
Air humidity:	up to the relative humidity† $95\%_R$@40°*C*

† To avoid vapor condensation

- time to the first ventilation[23]
- ventilation time at the controlled, maximum stream of the ventilating medium defined as time between the ventilation beginning[24] and the moment of the pressure drop to the level[25] of 1 *kPa*
- time to the next ventilation defined as the time from the moment when ventilation was finished until the carbon dioxide CO_2 partial pressure was to be increased to 1.5 *kPa*.

Based on the measured time to the first ventilation and Equation (6.5), it is possible to determine an average stream of carbon dioxide CO_2 emitted by the diver and finally to determine the time period to the next ventilation and the theoretical intensity of ventilation. In order to find practical ventilation, it is necessary to measure the ventilation time and the stream of the fresh ventilation air.

The calculation and measurements from manned diving results are presented as summarized data from the investigations in Table 6.15. The results from the manned dives are presented in Table 6.13 and Figure 6.8.

TABLE 6.13
The results of the measurements performed during dives with participation of the divers when interrupted ventilation was applied

		Diving 1		Diving 2	
№	Time	CO_2 contents	CO_2 partial pressure	CO_2 contents	CO_2 partial pressure
	t [*min*]	x_v [$m^3 \cdot m^{-3}$]	p_{CO_2} [*kPa*]	x_v [$m^3 \cdot m^{-3}$]	p_{CO_2} [*kPa*]
1	1.0	0.092	0.368	0.094	0.376
2	2.0	0.102	0.408	0.107	0.428
3	4.0	0.112	0.448	0.121	0.484
4	6.0	0.141	0.564	0.147	0.588
5	8.0	0.146	0.584	0.183	0.732
6	10.0	0.181	0.724	0.199	0.796
7	12.0	0.201	0.804	0.212	0.848
8	14.0	0.222	0.888	0.237	0.948
9	16.0	0.235	0.940	0.258	1.032
10	18.0	0.244	0.976	0.279	1.116
11	20.0	0.248	0.992	0.294	1.176
12	22.0	0.259	1.036	0.313	1.252
13	24.0	0.281	1.124	0.327	1.308
14	26.0	0.297	1.188	0.343	1.372
15	28.0	0.301	1.204	0.367	1.468
16	30.0	0.324	1.296	0.372	1.488
17	32.0	0.339	1.356	0.343	1.372
18	34.0	0.354	1.416	0.249	0.996
19	36.0	0.375	1.500	0.293	1.172
20	38.0	0.261	1.044	0.348	1.392
21	40.0	0.292	1.168	0.355	1.420
22	42.0	0.304	1.216	0.368	1.472
23	44.0	0.327	1.308	0.391	1.564
24	46.0	0.337	1.348	0.395	1.580
25	48.0	0.375	1.500		
26	50.0	0.369	1.476		

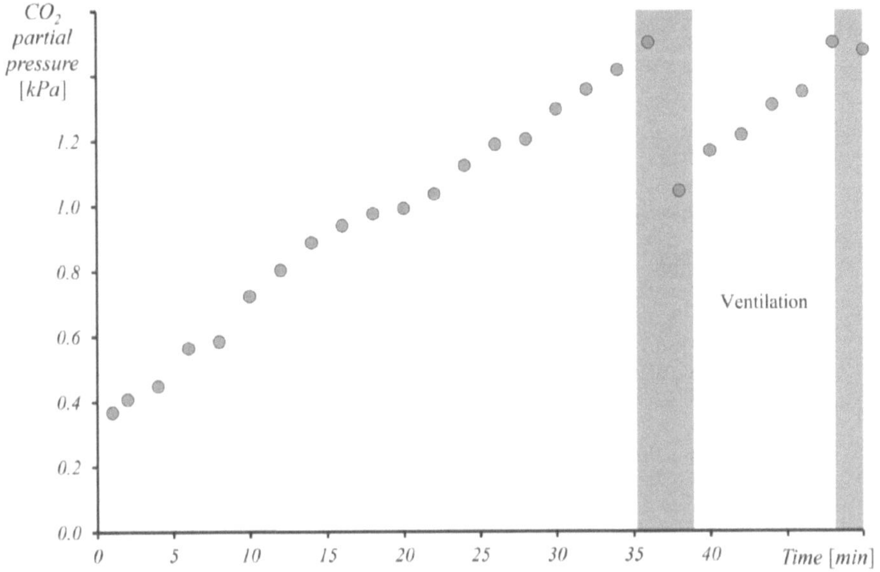

FIGURE 6.8 The example results of the measurements made during two manned hyperbaric expositions to interrupted ventilation – Diving N° 1 from Table 6.13.

The unmanned investigations were performed with the use of a carbon dioxide simulator. The simulator design has been described earlier. In this case, as distinct from the experiments with divers, the carbon dioxide CO_2 emission was known prior to the start of the experiments. The results of these experiments are presented in Table 6.14 and Figure 6.9.

Mechanical mixing of the chamber atmosphere[26] was an additional difference in the experiments as compared to those performed with the carbon dioxide CO_2 emission simulator. The difference affected the time necessary for lowering the carbon dioxide CO_2 contents in the chamber atmosphere and the necessary fresh ventilation medium demand during interrupted ventilation. Time necessary to decrease the carbon dioxide CO_2 contents in the chamber atmosphere during the interrupted ventilation experiments without divers[27] was significantly greater.[28]

As is evident from the Table 6.15, times to the first and next ventilation calculated are consistent with the experimental results, whereas the calculated demand for the fresh ventilation medium during the interrupted ventilation is 40–60% higher than that calculated for the mechanical mixing of the chamber atmosphere; without mixing, the gas demand is higher.

INVESTIGATIONS ON CONTINUOUS VENTILATION

There is no possibility to perform preliminary testing of assumed mathematical models during simulation dives with subjects, as it was in the previous experiments. The consistence of the assumed theoretical model can be tested using the carbon dioxide simulator that has been presented already. Two unmanned dives were performed. Their parameters are presented in Table 6.16.

TABLE 6.14

The results of the measurements made during unmanned dives when the carbon dioxide emission simulator was used and the interrupted ventilation was applied

		Diving 1		Diving 2	
№	Time t $[min]$	CO_2 contents x_v $[m^3 \cdot m^{-3}]$	CO_2 partial pressure p_{CO_2} $[kPa]$	CO_2 contents x_v $[m^3 \cdot m^{-3}]$	CO_2 partial pressure p_{CO_2} $[kPa]$
1	1.0	0.025	0.100	0.030	0.120
2	2.0	0.025	0.100	0.026	0.104
3	3.0	0.025	0.100	0.026	0.104
4	4.0	0.026	0.104	0.030	0.120
5	5.0	0.030	0.120	0.027	0.108
6	6.0	0.036	0.144	0.046	0.184
7	7.0	0.044	0.176	0.055	0.220
8	8.0	0.051	0.204	0.065	0.260
9	9.0	0.063	0.252	0.071	0.284
10	10.0	0.076	0.304	0.080	0.320
11	11.0	0.089	0.356	0.095	0.380
12	12.0	0.096	0.384	0.105	0.420
13	13.0	0.115	0.460	0.112	0.448
14	14.0	0.115	0.460	0.122	0.488
15	15.0	0.150	0.600	0.133	0.532
16	16.0	0.163	0.652	0.147	0.588
17	17.0	0.177	0.708	0.156	0.624
18	18.0	0.175	0.700	0.165	0.660
19	19.0	0.192	0.768	0.176	0.704
20	20.0	0.202	0.808	0.188	0.752
21	21.0	0.204	0.816	0.195	0.780
22	22.0	0.217	0.868	0.204	0.816
23	23.0	0.226	0.904	0.215	0.860
24	24.0	0.238	0.952	0.226	0.904
25	25.0	0.248	0.992	0.231	0.924
26	26.0	0.259	1.036	0.240	0.960
27	27.0	0.267	1.068	0.250	1.000
28	28.0	0.272	1.088	0.257	1.028
29	29.0	0.282	1.128	0.269	1.076
30	30.0	0.293	1.172	0.280	1.120
31	31.0	0.300	1.200	0.291	1.164
32	32.0	0.304	1.216	0.298	1.192
33	33.0	0.314	1.256	0.307	1.228
34	34.0	0.326	1.304	0.318	1.272
35	35.0	0.328	1.312	0.322	1.288
36	36.0	0.338	1.352	0.331	1.324
37	37.0	0.345	1.380	0.340	1.360
38	38.0	0.351	1.404	0.347	1.388
39	39.0	0.360	1.440	0.362	1.448
40	40.0	0.370	1.480	0.366	1.464
41	41.0	0.373	1.492	0.376	1.504
42	42.0	0.379	1.516	0.377	1.508
43	43.0	0.388	1.552	0.387	1.548
44	44.0	0.360	1.440	0.300	1.200

(Continued)

TABLE 6.14 ***(Continued)***

		Diving 1		Diving 2	
45	45.0	0.335	1.340	0.275	1.100
46	46.0	0.250	1.000	0.250	1.000
47	47.0	0.200	0.800	0.215	0.860
48	48.0	0.208	0.832	0.237	0.948
49	49.0	0.220	0.880	0.240	0.960
50	50.0	0.250	1.000	0.243	0.972
51	51.0	0.259	1.036	0.252	1.008
52	52.0	0.269	1.076	0.261	1.044
53	53.0	0.276	1.104	0.273	1.092
54	54.0	0.283	1.132	0.281	1.124
55	55.0	0.295	1.180	0.291	1.164
56	56.0	0.300	1.200	0.299	1.196
57	57.0	0.320	1.280	0.307	1.228
58	58.0	0.321	1.284	0.314	1.256
59	59.0	0.329	1.316	0.333	1.332
60	60.0	0.335	1.340	0.335	1.340
61	61.0	0.350	1.400	0.342	1.368
62	62.0	0.353	1.412	0.353	1.412
63	63.0	0.365	1.460	0.366	1.464
64	64.0	0.370	1.480	0.374	1.496
65	65.0	0.380	1.520	0.377	1.508
66	66.0	0.365	1.460	0.365	1.460
67	67.0	0.370	1.480	0.370	1.480
68	68.0	0.380	1.520	0.380	1.520

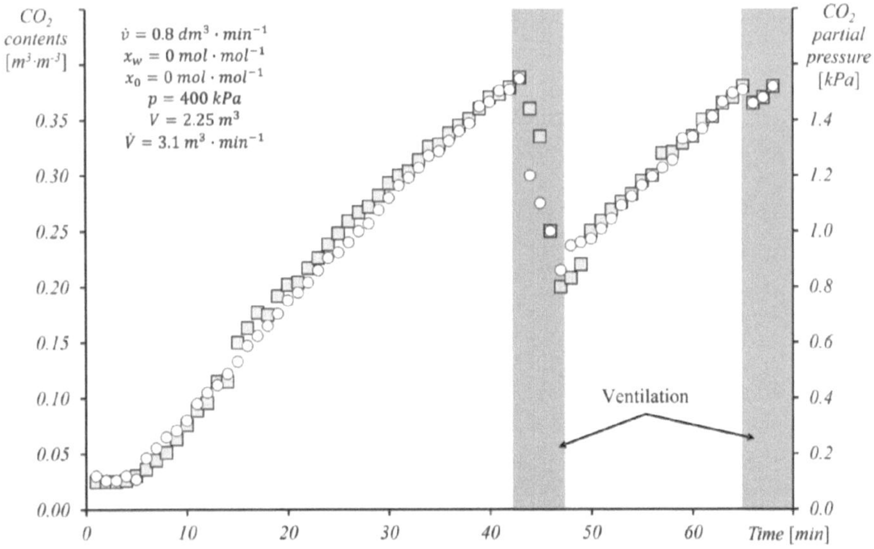

FIGURE 6.9 The experimental results obtained during unmanned dives with the carbon dioxide simulator and the interrupted ventilation applied.

TABLE 6.15

The measurement and calculation results based on diving with participation of the divers (interrupted ventilation was applied)

The number of diving	Theoretical values			The measured values					
	Time to the first ventilation	Ventilation air demand	Time to the next ventilation	Minimum stable value of CO_2 partial pressure after the first ventilation	Calculated average CO_2 emission per the diver	The measured time to the first ventilation	The measured ventilation time at maximum stream of ventilation gas	Calculated ventilation air demand	The measured time to the next ventilation
	$[min]$	$[m^3]$	$[min]$	$[kPa]$	$[dm^3 \cdot min^{-1}]$	$[min]$	$[min]$	$[m^3]$	$[min]$
1†	35.9	2.791	9.6	1.1	0.47	35.2	1.25•	3.88•	9.4
2†	31.3	3.649	10.4	1.0	0.54	31.2	1.9•	5.89•	10.2
1‡	42.2	5.112	18.3	1.0	0.4*	42.0	5.0#	15.5#	18.0
2‡	42.2	5.112	18.3	1.0	0.4*	42.0	5.1#	15.8#	18.0

Data: $x_0 = 0\ m^3 \cdot m^{-3}$; $x_w = 0\ m^3 \cdot m^{-3}\ p_{CO_2} = 1.5\ kPa$, $V = 2.25\ m^3$, $k = 2$ persons, $\dot{V} = 3.1\ m^3 \cdot min^{-1}$, $H = 30\ mH_2O$

† Diving with participation of people

‡ Diving with use of CO_2 emission simulator

* Emission fixed at the simulator

• The values obtained at mechanical mixing of the chamber atmosphere

The values obtained without mechanical mixing of the chamber atmosphere

TABLE 6.16

The results of the measurements made during unmanned dives when the carbon dioxide emission simulator was used and the continuous ventilation was applied

				Diving 1		Diving 2		Theoretical data	
The depth	The pressure	Time	The ventilation stream	Carbon dioxide contents	Carbon dioxide pressure	Carbon dioxide contents	Carbon dioxide pressure	Carbon dioxide contents	Carbon dioxide pressure
H [mH_2O]	p [kPa]	t [min]	V [$dm^3 \cdot min^{-3}$]	x_v [$\%_v$]	p_{CO_2} [kPa]	x_v [$\%_v$]	p_{CO_2} [kPa]	x_v [$\%_v$]	p_{CO_2} [kPa]
30	400	0	250	0.035	0.140	0.035	0.140	0.120	0.120
30	400	1	250					0.155	0.920
30	400	2	250	0.048	0.192			0.189	1.720
30	400	3	250			0.034	0.136	0.222	2.520
30	400	4	250					0.255	3.320
30	400	5	250			0.026	0.104	0.286	4.120
30	400	6	250					0.317	4.920
30	400	7	250			0.042	0.168	0.346	5.720
30	400	8	250	0.097	0.388			0.375	6.520
30	400	9	250					0.403	7.320
30	400	10	250			0.061	0.244	0.430	8.120
30	400	11	250					0.457	8.920
30	400	12	250					0.483	9.720
30	400	13	250	0.142	0.568			0.508	10.520
30	400	14	250					0.532	11.320
30	400	15	250			0.112	0.448	0.556	12.120
30	400	16	250					0.579	12.920
30	400	17	250					0.602	13.720
30	400	18	250	0.170	0.680			0.624	14.520
30	400	19	250					0.645	15.320
30	400	20	250			0.149	0.596	0.666	16.120
30	400	21	250					0.686	16.920
30	400	22	250					0.705	17.720
30	400	23	250	0.188	0.752			0.724	18.520
30	400	24	250					0.743	19.320
30	400	25	250			0.176	0.704	0.761	20.120
30	400	26	250					0.778	20.920
30	400	27	250					0.795	21.720
30	400	28	250	0.220	0.880			0.812	22.520
30	400	29	250					0.828	23.320
30	400	30	250			0.200	0.800	0.844	24.120
30	400	31	250					0.859	24.920
30	400	32	250					0.874	25.720
30	400	33	250	0.236	0.944			0.888	26.520
30	400	34	250					0.902	27.320
30	400	35	250			0.224	0.896	0.916	28.120
30	400	36	250					0.929	28.920
30	400	37	250					0.942	29.720
30	400	38	250	0.251	1.004			0.955	30.520
30	400	39	250					0.967	31.320
30	400	40	250			0.237	0.948	0.979	32.120
30	400	41	250					0.990	32.920
30	400	42	250					1.001	33.720
30	400	43	250	0.265	1.060			1.012	34.520

(Continued)

TABLE 6.16 ***(Continued)***

				Diving 1		Diving 2		Theoretical data	
The depth	The pressure	Time	The ventilation stream	Carbon dioxide contents	Carbon dioxide pressure	Carbon dioxide contents	Carbon dioxide pressure	Carbon dioxide contents	Carbon dioxide pressure
30	400	44	250					1.023	35.320
30	400	45	250			0.253	1.012	1.033	36.120
30	400	46	250					1.043	36.920
30	400	47	250					1.053	37.720
30	400	48	250	0.277	1.108			1.063	38.520
30	400	49	250					1.072	39.320
30	400	50	250			0.265	1.060	1.081	40.120
30	400	51	250					1.090	40.920
30	400	52	250					1.098	41.720
30	400	53	250	0.287	1.148			1.106	42.520
30	400	54	250					1.114	43.320
30	400	55	250			0.288	1.152	1.122	44.120
30	400	56	250					1.130	44.920
30	400	57	250					1.137	45.720
30	400	58	250	0.296	1.184			1.144	46.520
30	400	59	250					1.151	47.320
30	400	60	250			0.296	1.184	1.158	48.120
30	400	61	250					1.165	48.920
30	400	62	250					1.171	49.720
30	400	63	250	0.304	1.216			1.178	50.520
30	400	64	250					1.184	51.320
30	400	65	250			0.306	1.224	1.190	52.120
30	400	66	250					1.195	52.920
30	400	67	250					1.201	53.720
30	400	68	250	0.313	1.252			1.206	54.520
30	400	69	250					1.212	55.320
30	400	70	250			0.317	1.268	1.217	56.120
30	400	71	250					1.222	56.920
30	400	72	250					1.227	57.720
30	400	73	250	0.316	1.264			1.232	58.520
30	400	74	250					1.236	59.320
30	400	75	250			0.323	1.292	1.241	60.120
30	400	76	250					1.245	60.920
30	400	77	250					1.249	61.720
30	400	78	250	0.321	1.284			1.253	62.520
30	400	79	250					1.257	63.320
30	400	80	250			0.323	1.292	1.261	64.120
30	400	81	250					1.265	64.920
30	400	82	250					1.269	65.720
30	400	83	250	0.318	1.272			1.272	66.520
30	400	84	250					1.276	67.320
30	400	85	250			0.329	1.316	1.279	68.120
30	400	86	250					1.283	68.920
30	400	87	250					1.286	69.720
30	400	88	250	0.332	1.328			1.289	70.520
30	400	89	250					1.292	71.320
30	400	90	250			0.335	1.340	1.295	72.120
30	400	91	250					1.298	72.920
30	400	92	250					1.301	73.720

(Continued)

TABLE 6.16 *(Continued)*

				Diving 1		Diving 2		Theoretical data	
The depth	The pressure	Time	The ventilation stream	Carbon dioxide contents	Carbon dioxide pressure	Carbon dioxide contents	Carbon dioxide pressure	Carbon dioxide contents	Carbon dioxide pressure
30	400	93	250	0.329	1.316			1.303	74.520
30	400	94	250					1.306	75.320
30	400	95	250			0.342	1.368	1.309	76.120
30	400	96	250					1.311	76.920
30	400	97	250					1.313	77.720
30	400	98	250	0.331	1.324			1.316	78.520
30	400	99	250					1.318	79.320
30	400	100	250			0.353	1.412	1.320	80.120
30	400	101	250					1.323	80.920
30	400	102	250					1.325	81.720
30	400	103	250	0.335	1.340			1.327	82.520
30	400	104	250					1.329	83.320
30	400	105	250			0.350	1.400	1.331	84.120
30	400	106	250					1.333	84.920
30	400	107	250					1.334	85.720
30	400	108	250	0.330	1.320			1.336	86.520
30	400	109	250					1.338	87.320
30	400	110	250			0.361	1.444	1.340	88.120
30	400	111	250					1.341	88.920
30	400	112	250					1.343	89.720
30	400	113	250	0.334	1.336			1.345	90.520
30	400	114	250					1.346	91.320
30	400	115	250			0.366	1.464	1.348	92.120
30	400	116	250					1.349	92.920
30	400	117	250					1.350	93.720
30	400	118	250	0.338	1.352			1.352	94.520
30	400	119	250					1.353	95.320
30	400	120	250			0.358	1.432	1.354	96.120
30	400	121	250					1.356	96.920
30	400	122	250					1.357	97.720
30	400	123	250	0.327	1.308			1.358	98.520
30	400	124	250					1.359	99.320
30	400	125	250			0.355	1.420	1.360	100.120
30	400	126	250					1.361	100.920
30	400	127	250					1.362	101.720
30	400	128	250	0.334	1.336			1.363	102.520
30	400	129	250					1.364	103.320
30	400	130	250			0.350	1.400	1.365	104.120
30	400	131	250					1.366	104.920
30	400	132	250					1.367	105.720
30	400	133	250	0.331	1.324			1.368	106.520

The comparison of the experimental and theoretical results is given in Figure 6.10 and Table 6.17. Worthy of note is that the theoretical results obtained are consistent with the results obtained in the experiments carried out in real circumstances.

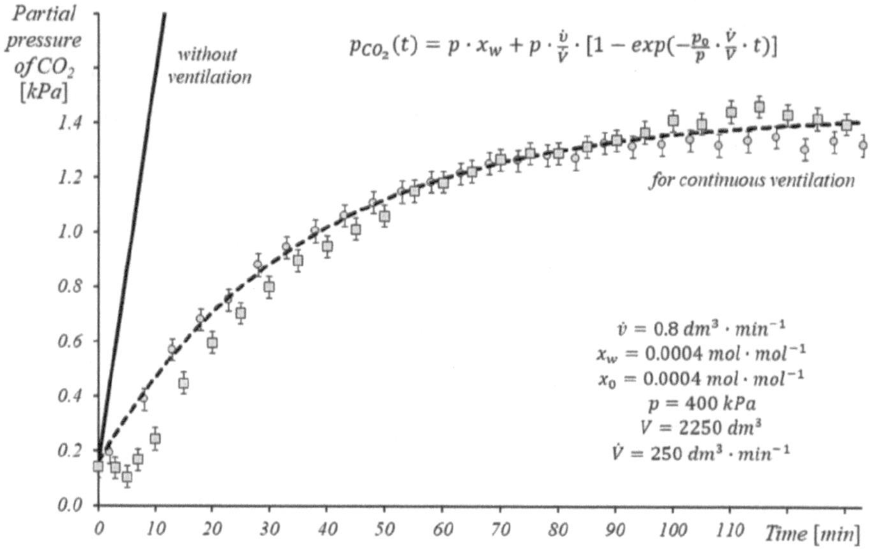

FIGURE 6.10 The comparison of experimental and theoretical results of continuous ventilation of the diving system, type ORTOLAN L-80.

TABLE 6.17

The conditions of experimental measurements of continuous ventilation effectiveness with the use the simulator for the controlled carbon dioxide emission

The number of diving	The depth	Approximated pressure corresponding the depth	Minimum stream of the ventilation medium	CO_{22} emission
	H	p	$\dot{V}$	$\dot{v}$
	$[mH_2O]$	$[kPa]$	$[dm^3 \cdot min^{-1}]$	
1	30	400	250	0.8
2	30	400	250	0.8

$x_0 = 0.0004\ m^3 \cdot m^{-3}$; $x_w = 0.0004\ m^3 \cdot m^{-3}$; $V = 2.25\ m^3$
Rotameter: 50–400 ± 20 $dm^3 \cdot min^{-1}$

SUMMARY

The results of the preliminary experiments confirmed the usability of the applied mathematical models for the interrupted and continuous ventilation parameters. The earlier assumptions concerning the necessity to use undetermined ventilating medium excess were confirmed as well. As it follows from the above, in order to find the mathematical model that can be used to calculate the maximum excess of ventilating medium, it is necessary to carry out further experiments.

The aim of the presented experiments was to find out whether the ventilation mathematical models applied to the semi-closed-circuit diving apparatus SCR can be applied also to hyperbaric chambers. The theoretical formulae for the interrupted and continuous ventilation of the hyperbaric facilities are presented. The experimental results of ventilation effectiveness of the hyperbaric chamber, type ORTOLAN L-80, are presented as well. The results of the preliminary experiments confirmed the usability of the applied mathematical models for the interrupted and continuous ventilation parameters. The experiments confirmed the possibility of using the common ventilation mathematical model for two very different hyperbaric facilities such as the diving apparatus and the hyperbaric chamber.

The procedures proposed to develop mathematical models of ventilation can be applied on a micro as well as the macro scale. The macro scale is concerned with the diving methodology. It is important that the empirical and semi-empirical models, when compared to the presented analytical models, are related to the physical phenomena occurring in the hyperbaric environment.

The mathematical models of the hyperbaric chamber ventilation presented in the literature are not useful, as they are incorrect. They:

- are not consistent with the ventilation tables calculated (Haux, 1982)
- do not fulfill the boundary conditions (Reimers & Hansen, 1972).

NOTES

1 Chambers or the diving complexes.
2 E.g., after rapid compression.
3 E.g., in order to modify the composition of the hyperbaric complex atmosphere.
4 The last two parameters can be evaluated on the basis of divers' subjective feelings.
5 Or the stream of the atmosphere removing from the hyperbaric facility for keeping stable pressure in the hyperbaric chamber.
6 With an external gas supply and gas exhalation outside the hyperbaric facility.
7 ***B**uilt-**i**n-**B**reathing **S**ystem.*
8 Joule–Thomson effect
9 Hypercapnia.
10 It was assumed that carbon dioxide emission is equal to the oxygen consumption.
11 For one person $k = 1$.
12 With the use of an external regeneration systems of the breathing medium.
13 Despite the fact that it is physiologically more suitable.
14 The problems of maintaining the pressure at the proper level, difficulties with ventilation at simultaneous decompression process, etc.
15 The uniform, tiresome noise of the ventilation medium flowing through the installation.
16 The choice was caused by the *MVB* available.
17 The calculations performed by means described earlier.
18 The number of turns.
19 At the diving depths of 10, 20, and 30 mH_2O, the pressure was increased by 1.5 mH_2O and at the fixed valve position decrease pressure until the moment when the depth decreases by 3 mH_2O. During this operation the inlet valve is closed.
20 Calibrated.
21 Internal pressure in the chamber.
22 The choice of the diving depth was made by using the worst-circumstances principle.

23 Is defined as time from the beginning of the pressure increase until the carbon dioxide CO_2 partial pressure at the equivalent depth of 30 mH_2O reaches 1.5 kPa.
24 Carbon dioxide CO_2 partial pressure in the chamber reached the level of 1.5 kPa.
25 It is necessary to determine the stable level of the carbon dioxide CO_2 partial pressure that was lowered during ventilation.
26 The divers mixed the chamber atmosphere by flapping towels.
27 Without mixing of the chamber atmosphere.
28 Over 2.5 times longer than with mixing of the chamber atmosphere.

REFERENCES

Goliński J. A. & Troskalański A. T. 1979. *Strumienice*. Warszawa: WNT.

Haux G. 1982. *Subsea manned engineering*. London: Baillière Tindall. ISBN 0-7020-0749-8.

Kłos R. 2000. *Aparaty Nurkowe z regeneracją czynnika oddechowego*. Poznań: COOPgraf. ISBN 83-909187-2-2.

Kłos R. 2007a. Modelowanie procesów wentylancji obiektów normo- i hiperbarycznych. Higher PhD diss., The Polish Naval Academy. PL ISSN 0860-889X nr 160A.

Kłos R. 2007b. *Mathematical modelling of the normobaric and hyperbaric facilities ventilation*. Gdynia: Polish Hyperbaric Medicine and Technology Society. ISBN 978-83-924989-0-2.

Kłos R. 2012. *Możliwości doboru ekspozycji tlenowo-nitroksowych dla aparatu nurkowego typu AMPHORA – założenia do nurkowań standardowych i eksperymentalnych*. Gdynia: Polish Hyperbaric Medicine and Technology Society. ISBN 978-83-924989-8-8.

Kłos R. 2014. *Helioksowe nurkowania saturowane – podstawy teoretyczne do prowadzenia nurkowań i szkolenia*. Gdynia: Polish Hyperbaric Medicine and Technology Society. ISBN 978-83-938322-1-7.

Kłos R. 2015. *Katalityczne utlenianie wodoru na okręcie podwodnym*. Gdynia: Polish Hyperbaric Medicine and Technology Society. ISBN 978-83-938-322-3-1.

Mittleman J. 1989. Computer modeling of underwater breathing systems. In: *Physiological and human engineering aspects of underwater breathing apparatus*, ed. D. E. Warkander & C. E. G. Lundgren, 223–270. Bethesda: Undersea and Hyperbaric Medical Society.

Przylipiak M. & Torbus J. 1981. *Sprzęt i prace nurkowe-poradnik*. Warszawa: Wydawnictwo Ministerstwa Obrony Narodowej. ISBN 83-11-06590-X.

Reimers S. D. & Hansen O. R. 1972. *Environmental control for hiperbaric applications*. Panama City: US Navy Experimental Diving Unit. NEDU Rep. 25-72.

US Navy Diving Manual. 1980. Carson: Best Publishing Co. NAVSEA 0994-LP-001-9010.

US Navy Diving Manual. 2016. The Direction of Commander: Naval Sea Systems Command. SS521-AG-PRO-010 0910-LP-115-1921.

Part III

Submarines

Part III describes the research on the ventilation of submarines.

Most of the compartments of submarines are much larger than the compartments of hyperbaric complexes. In the research on the ventilation of submarine compartments the ventilation agent was released into the atmosphere. In order for the ventilation process to take place in the submarine's compartment, the pressure inside it had to be increased. Thus, the submarine compartment can be regarded as an object with the pressure slightly higher than the atmospheric pressure.

7 Submarine Atmosphere Monitoring System

Use of an appliance to monitor the breathing gas with a computer-based data acquisition and data analysis system is a standard not only in submarines. Maintaining the proper composition of submarine atmosphere is essential for a mission and crew safety. The most important parameters of submarine atmosphere should be monitored on a continuous basis. Apart from providing life support in the confined space, it enables life support systems to work in an automatic mode, and it is employed to evaluate hazards to naval facilities.

Implementation of new gas analyzers requires special fitting procedures. The number of laboratory tests necessary to commence the implementation research phase may be different and may be reduced if the gas analyzer has already been tested and is now only being modified or certified by an acknowledged laboratory. In the case of a new gas analyzer, the number of laboratory tests must be sufficient to reach the proper approved confidence level.

Monitoring atmosphere in submarines is very important, particularly in an emergency situation. A decision to abandon a distressed submarine (DISSUB) is directly dependent on the actual state of the atmosphere. During normal duty, readouts from the monitored atmosphere system are very useful in many situations, e.g., start/stop of the ventilation/regeneration system, flushing battery hole, hydrogen burning system, possibility of entering compartments with weak ventilation, and so on.

In NATO, the minimum requirements for atmospheric monitoring equipment to be available in all submarine inhabitable compartments was defined in STANG 1320 (NATO Standardization Agreement STANAG 1320 UD, 2005). This document and additional requirements were used to develop the minimum requirements for building an atmosphere monitoring system in the KILO-class submarine showed in Table 7.1.

IMPLEMENTATION OF NEW GAS ANALYZERS

The systems originally used could not provide continuous and simultaneous local measurements at different measuring points. The new system was based on a centrally located measuring stand, where gas suction was performed with the aid of a pipe system.

The advantage of the local measuring system is that measurements are achieved at the stand where they are necessary. Only analytic signals are transmitted to the central computer. The computer data acquisition system enables analyzing the situation. Obviously, data transmission is considerably faster than sampling gas, but the modernization of the measuring system does not have to be directed toward elimination of the existing measuring pipe system. In this case a mixed system could be used.

TABLE 7.1
The preliminary minimum requirements for atmospheric monitoring equipment located in submarines

BUILT-IN-SYSTEM

Type of measurement	Range	Additional requirements
atmospheric overpressure	0–600 ± 1 *kPa*	should be measured continuously
percentage content of O_2	0–25 ± 0.2$\%_v$	
percentage content of CO_2	0–4.5 ± 0.01$\%_v$	
percentage content of H_2	0–100 ± 1% *LEL*	
temperature	−40–60°*C* ± 1% (*relative*)	
relative humidity	0–100 ± 5% (*relative*)	

PORTABLE SYSTEM

Type of measurement	Range	Additional requirements
percentage content of O_2	0–25 ± 0.2$\%_v$	may be measured at intervals
percentage content of CO_2	0–4.5 ± 0.01$\%_v$	
percentage content of H_2	0–100 ± 1% *LEL*	
percentage content of *CO*	0–200 *ppm*	

All equipment must be robust, saltwater resistant, simple to operate, and accurate at absolute pressures of up to 0.6 *MPa*. The equipment must be able to withstand temperatures running from −2°*C* to +50°*C* and must be independent of external energy sources.
Measurements should be recalculated into other metric units by a computerized system.

Implementation of a built-in-submarine monitoring system with local measurements became possible as gas analyzers became cheaper and able to work for longer periods without engineering supervision, e.g., tuning, readjusting, calibration, and so on.

Implementation of gas analyzers requires that special fitting procedures be followed. The flowchart showing implementation of the new type of gas analyzers is presented in Figure 7.1.

The number of laboratory tests required to carry out in order to commence the implementation research phase may be different and may be reduced if the gas analyzer has already been tested and now is only being modified, or has been certified, by an approved laboratory. In the case of a new gas analyzer, the number of laboratory tests must be sufficient to reach the proper confidence level.

Decision-makers from the ministry of defense make a decision to terminate the laboratory research phase and to commence the implementation research phase.

The implementation research phase involves observation of the measuring system under working conditions. During the implementation research, gas analyzers are only tested. In order to control the submarine atmosphere, other gas analyzers are used.

The implementation stage may be stopped at any moment, whenever some problems occur. If there are no problems during the implementation phase, decision-makers give an approval to use the gas analyzers. However, the research is not finished yet.

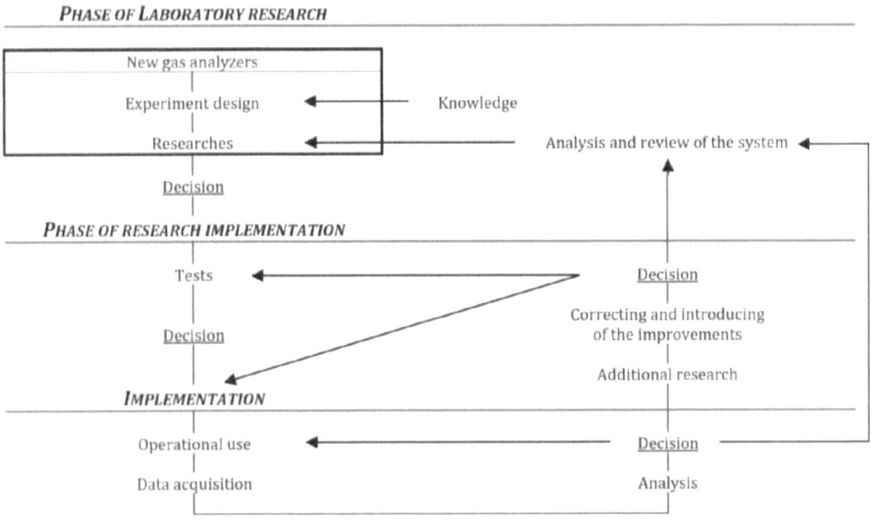

FIGURE 7.1 The flowchart of implementation of the new type of gas analyzers in the Polish Navy.

During a normal operational use, the results obtained from the gas analyzers are periodically analyzed by a specialist from the navy. If some trouble with the use of the system occurs, the implementation phase must be repeated. When it is decided to stop using or modernize the measuring system, the whole research process starts from the beginning.

LABORATORY RESEARCH

Laboratory research was carried out within the framework of a project financially supported by the State Committee for Scientific Research and the Polish Navy. That stage of the project includes selection of gas analyzers[1] fulfilling the technical requirements set by the navy and laboratory research to check the technical requirements presented in Table 7.1.

Gas analyzers should fulfill the technical requirements set for equipment on navigation bridges and in the marine power plant. Hence, in regard to work site environmental properties, all the above-mentioned equipment should fulfill the requirements of the acknowledged marine classification societies. Maintenance should be carried out at intervals not longer than 6 months.

All the equipment must keep the working parameters under special environmental conditions, i.e., pressure of 80–600 *kPa* and tilt of 30°(3 *min*)/45°(9 *s*). It must be characterized by resistance to disturbances in the working environment caused by chemical contaminants such as oxygen 19–25 $\%_v$, hydrogen $<4\%_v$, carbon dioxide $<1.5\%_v$, carbon monoxide $<15\ mg \cdot m^{-3}$, hydrocarbons $<100\ mg \cdot m^{-3}$, antimony hydride $<0.15\ mg \cdot m^{-3}$, Freon-12 $<500\ mg \cdot m^{-3}$, sulphur dioxide $<0.4\ mg \cdot m^{-3}$, chloroprene dimers $<1.5\ mg \cdot m^{-3}$, nitrogen dioxide $<0.5\ mg \cdot m^{-3}$, hydrogen sulphide $<0.5\ mg \cdot m^{-3}$, and ammonia $<0.8\ mg \cdot m^{-3}$.

FIGURE 7.2 Computer test stand to conduct research on selected metrology features of gas analyzers, general view: 1 – climatic chamber for environmental measurements, 2 – control-measuring desk, 3 – monitor, 4 – control computer, 5 – power pack with battery support, 6 – electric control box equipped with measuring transducers.

The research on gas analyzers was conducted on the computer test stand that had been built for measurements of basic metrology features (Figure 7.2).

The control desk of the research stand was equipped with computer-controlled rapid electromagnetic valves and flow regulators that were used to control the standard flow of gases. A system of signal and controllable transducers was placed inside an electric control box.

The test stand was also equipped with a computer-controlled climatic chamber. The following parameters were measured on the test stand: discrimination threshold, dead time, reaction and response time, reproducibility of measurements, reproducibility of scaling and adjusting,[2] reversibility of indications,[3] effect of high and low temperatures, long-term deviation, drift, reaction on some contaminants of the monitored atmosphere,[4] influence of other physical quantities, frequency of metrological certification, etc. Experiments were also performed under increased pressure in the hyperbaric diving complex and during simulation of the system operation in a manned submersible (Figure 7.3).

Different types of gas analyzers from various manufacturers were investigated. The experiments were limited to analyzing technical parameters and basic metrology features of the gas analyzers selected. Special emphasis was put on the situations where the test results revealed that the permissible parameters specified by the manufacturer or in the assumptions that had been made were exceeded.

(a)

(b)

FIGURE 7.3 Experimental diving complex at the Naval Academy

(Continued)

(c)

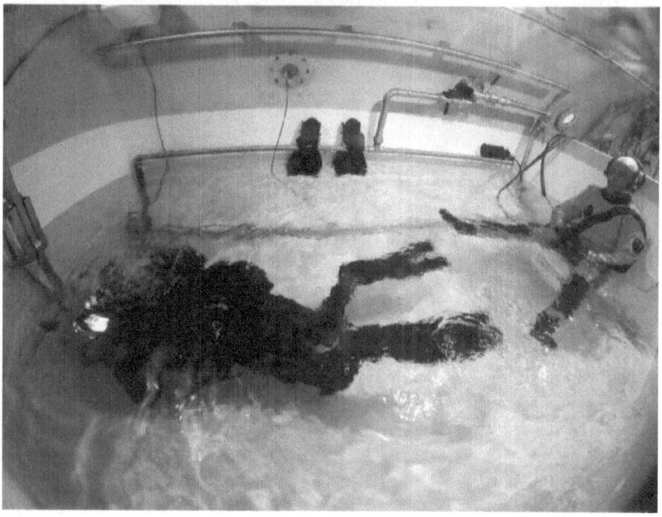

(d)

FIGURE 7.3 (Continued) Experimental diving complex at the Naval Academy.

This procedure, laboratory study, results of experiments, and chosen analyzers built into a submarine monitoring system were described in detail elsewhere (Kłos, 2008; Kłos, 2015). This research created a base for building an experimental submarine atmosphere monitoring system. Only the preliminary temperature and pressure test results for three types of gas analyzers used to make a built-in submarine system are presented: DrägerSensorO2LS, DrägerSensorO2, and DrägerPolytron IR CO2.

The gas analyzers were tested with the respect to their basic metrology features, i.e., long-term reading out of the stable parameters of the reference atmosphere, temperature tests, reaction time, measuring error in use of 5÷7 sets of standard samples, drift, discrimination threshold, dead time, hysteresis, repeatability of the measurements, correctness of calibration and adjustment procedures, effect of pressure and other parameters, resistance for disturbances resulting from contaminants[5] that may occur in the atmosphere of a submersible. Some of the above-mentioned parameters were investigated in the preliminary phase.

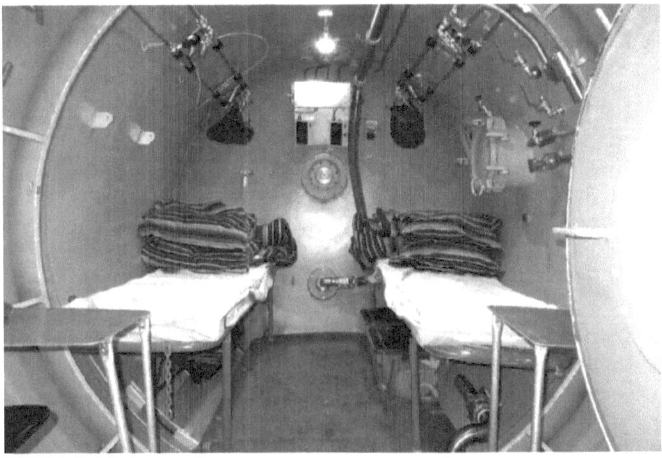

(e)

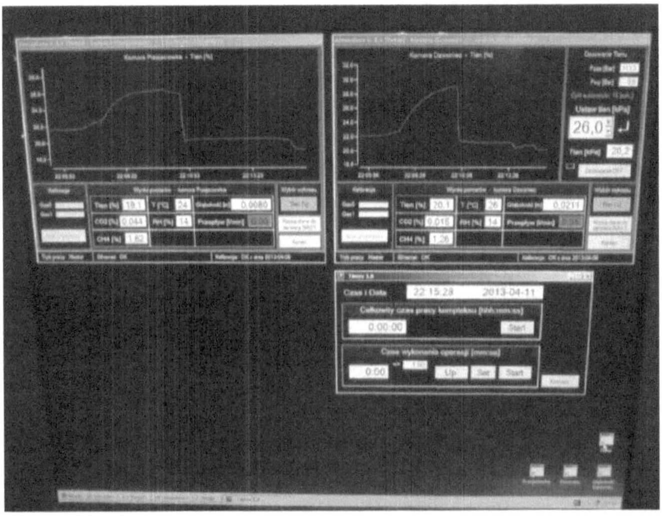

(f)

FIGURE 7.3 (Continued) Experimental diving complex at the Naval Academy.

Functioning of the sensors in the phase of the laboratory research experiments phase at various temperatures and pressures will be described with regard to the situations where the test results revealed that the permissible parameters specified by the manufacturer were exceeded.

THE PRELIMINARY TEST

The aim of the test was to measure the sensors' stability indications and determine the effect of pressure and temperature on the results. The results were compared with the parameters provided by the manufacturer. In order to determine sensors' stability indications, a 150 *h* long test was performed. During the test, where the sampling time was 1 *min*, the following parameters were measured on the continuous basis:

- ambient temperature *t* [°*C*] with the aid of the Vaisala sensor of the type 50U/50Y - maximum limiting error ±0.8°*C*,
- atmospheric pressure p_a [*kPa*] with the aid of the Honeywell sensor of the type PPT - maximum limiting error ±0.2 *kPa*,

In order to design the measuring system, the investigated sensors were applied. The measuring set was placed inside a tank having $d_z \cong 2000$ *mm* in outside diameter and the length $L \cong 3751$ *mm*. In this way the measuring system was protected against atmospheric motion and instantaneous fluctuation of temperature, humidity, etc. During the test, gas analyzers types DrägerSensorO2 and DrägerSensorO2LS slightly exceeded the limiting value of systematic error specified by the manufacturer. The gas analyzer type DrägerPolytronIRCO2 did not show such deviations.

THE TEMPERATURE TESTS

The aim of the first kind of tests was to determine the effect of temperature variations on the indications of the gas analyzer under investigation. In order to determine the effect of temperature on the indications of the gas analyzer, the gas analyzer was placed inside the environmental chamber, which is a part of the computer stand presented in Figure 7.2.

The climatic chamber enables regulation of temperature within a temperature range[6] 15–40°*C* and measurement of relative humidity within a humidity range[7] 10–90$\%_{rh}$. Each of the sensors was investigated separately; the test duration was 7 *h*.

The tests showed that the indications of the investigated sensors of oxygen percentage contents are temperature dependent. In the case of oxygen sensors, the decrease in temperature influenced the indications of the sensors. The indications of the DrägerSensorO2LS did not exceed the permissible measuring error specified by the manufacturer. Only in the case of humidity variations were the permissible measuring errors negligibly exceeded.

In the case of the DrägerSensorO2, the permissible measuring error was negligibly exceeded when the temperature decreased (Figure 7.4).

However, the indications of the DrägerPolytronIR sensor did not exceed the permissible measuring errors. During faster variations in temperature, for example from 15°*C* to 30°*C* within a time range 15–40 *min*, the permissible measuring errors were not exceeded.[8]

The aim of the other tests was to determine the effect of the slow temperature changes on the indications of the investigated gas analyzer. The tests were performed with the use of the stand described earlier. The investigated gas analyzers were placed inside the climatic chamber, the mean temperature inside the chamber was maintained at a level of approximately 34°*C*. Duration of tests was over 100 *h*.

In the case of the other investigated sensors, the effect of changes in temperature and humidity on exceeding the permissible limits of errors specified by the

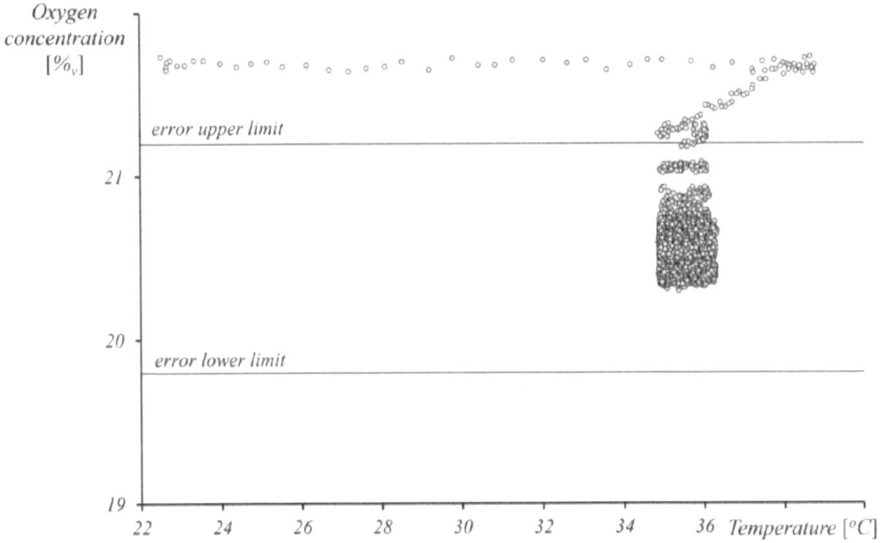

FIGURE 7.4 Function of oxygen concentration vs time $C_{O_2} = f(t)$ for DrägerSensorO_2.

manufacturer was not recorded. The indications of the DrägerSensorO2 did not exceed the assumed permissible deviations at the full range of humidity changes. The tests performed within a temperature range 5–7°*C* during approximately 50 *h* showed that the indications of the DrägerSensorO2LS negligibly exceeded the permissible measuring error, while the indications of the other sensors were correct.

THE PRESSURE TESTS

The aim of the tests was to determine the effect of pressure on indications of the gas analyzer. The gas analyzers were placed inside the hyperbaric complex, type DGKN-120 (Figure 7.3).

A series of expositions approximately 2 *h* long at the depths 10 mH_2O, 30 mH_2O, and 60 mH_2O were performed. During the experiments the pressure,[9] temperature,[10] and humidity were measured.

The indications of all the investigated sensors are dependent on the variable partial pressure of the measured quantities – for example, Figures 7.5 and Figure 7.6 present indications of the investigated sensors during pressure increase corresponding to the depth.

The DrägerSensorO2 is typical of membrane sensors. The principle of operation is reaction to partial pressure by measured gas concentration (Figure 7.5). DrägerSensor O2LS is equipped with a sufficiently long, small-diameter diffusion channel through which the gas can penetrate. For this reason it does not react to the oxygen partial pressure but to the percentage of oxygen. Hence, its indications are not dependent on the external pressure except for disturbances that occur at the beginning of the pressure increase. This leads to an increase in the measuring error.

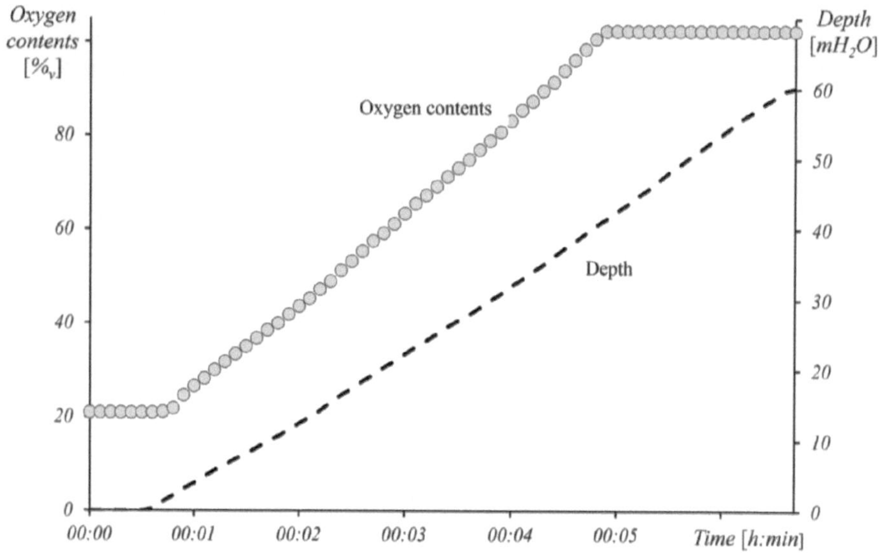

FIGURE 7.5 Indications of the DrägerSensorO2 during increase in pressure.

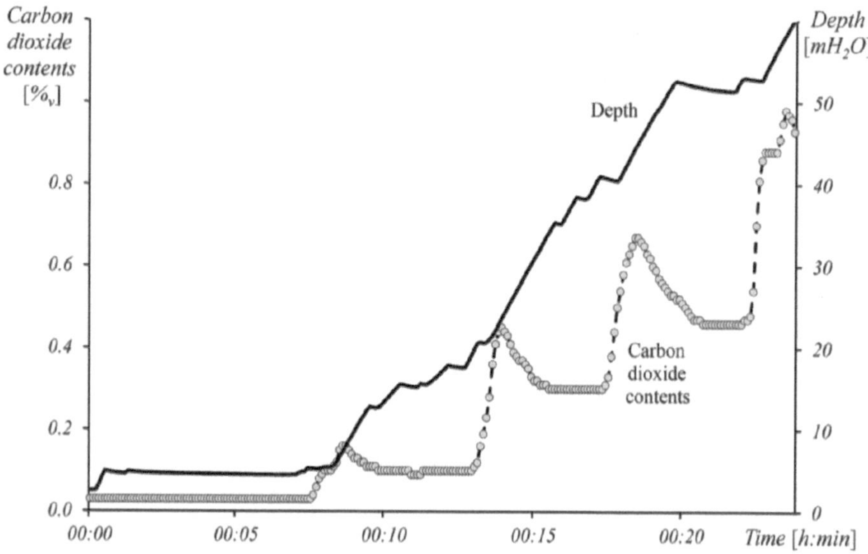

FIGURE 7.6 Indications of the DrägerPolytronIR CO2 sensor during an increase in chamber pressure.

The indications of the DrägerPolytronIRCO2 show significant deviations from the right contents of carbon dioxide in the investigated atmosphere (Figure 7.6). This is a typical phenomenon. With infrared technology, a change in component partial pressure or a change in total sample pressure can increase or weaken the intensity of the corresponding band without maintaining the proportionality to changes in gas concentration. An increase of pressure may lead to the so-called pressure band transfer. It causes poor resolution and false contour in the absorption band and finally incorrect results. Further research will be focused on this problem.

The results of the other experiments confirmed the parameters provided by the manufacturer. Later, the DrägerSensorO2LS was withdrawn from use because the hydrogen contents in the investigated atmosphere affected the gas analyzer accuracy. Based on the laboratory experimental results, the following gas analyzers were admitted to further research: DrägerSensorO2, Dräger PolytronIRCO2, and DrägerSensorH2. In order to prepare an experimental measuring system, the gas analyzers already tested were employed.

SUMMARY

A Polish Navy KILO-class submarine was equipped with an experimental submarine atmospheric monitoring system. This system was built by the Polish Naval Academy. The monitoring system consisted of oxygen, hydrogen, carbon dioxide gas analyzers, and temperature, moisture, and pressure sensors that met the minimum requirements for atmospheric monitoring equipment installed in submarines listed in Table 7.1.

The sensors and analyzers are connected to a central industrial computer by data acquisition modules into RS 485 standard net. Readouts can be recorded by a computer program, every second or longer period of time. Typical readout time is 5 *s*, and on a typical hard disk it is possible to collect and store data for at least 3 *years*.

Since May, 18, 2002, when the system was turned on, the system has been used to collect[11] readouts. Up to now, no malfunctions of the monitoring system have been recorded. In this case, it has been proved that RS 485 standard is robust enough for submarine applications.

NOTES

1 Based on the technical data.
2 Calibration repeatability.
3 Hysteresis.
4 Cross-sensitivity.
5 Metabolic gases, technical pollutants.
6 Temperature measurement with the aid of temperature sensor, type Vaisala Humiter 50U/50Y, maximum limiting error ±0.8°*C*.
7 Measurement with the aid of humidity sensor, Vaisala Humiter 50U/50Y, maximum relative error $\pm 1\%_{rh} rh$.
8 Observations up to 5 *h* after the change in temperature.
9 Pressure sensor of the type *Rosemount Alpha – line Pressure Transmitter* – maximum absolute error ±0.3 mH_2O.
10 Humidity/temperature sensor of the type *Vaisala* 50*U*/50*Y* – maximum absolute error ±0.8°*C*.
11 Every 5*s*, 24 *h* a day.

REFERENCES

Kłos R. 2008. *Systemy podtrzymania życia na okręcie podwodnym*. Gdynia: Polish Hyperbaric Medicine and Technology Society. ISBN 978-83-924989-4-0.

Kłos R. 2015. *Katalityczne utlenianie wodoru na okręcie podwodnym*. Gdynia: Polish Hyperbaric Medicine and Technology Society. ISBN 978-83-938-322-3-1.

NATO Standardization Agreement STANAG 1320 UD. 2005. *Minimum requirements for atmospheric monitoring equipment located in submarines with escape capability*. Brussels: NATO Standardization Office. NSO(NAVAL)1449(2014)SMER/1320.

8 Ventilation of Submarines

The portable laboratory, showed in Figure 8.1, was developed and equipped, and the Submarine Rescue Ship (SRS) air bank used to precisely measure pressure had to be prepared before distressed submarine (DISSUB) ventilation measurements could be started.

SUBMARINE ATMOSPHERIC MONITORING SYSTEM

A Polish KILO-class submarine was equipped with an experimental submarine atmospheric monitoring system. This system was built by the Naval Academy (Kłos, 2008). The monitoring system consists of oxygen, hydrogen, carbon dioxide gas analyzers, and temperature, moisture, and pressure sensors. The sensors and analyzers are connected to a central industrial computer by data acquisition modules into RS 485 net. Readouts can be recorded every second or longer period of time. Typical readout time is 5 s, and on a typical hard disk it is possible to collect and store data for at least 3 years (Figure 8.2).

SOFTWARE AND DATA TRANSMISSION

The submarine measuring net RS485 was connected by a deck-mounted telephone connector and a waterproof electric cable comprising a twisted pair of conductors connected to the portable laboratory, as well as via an RS485/RS232 converter to a typical PC, capable of monitoring atmosphere in the submarine using a special computer program.

METHOD

The test consisted of a simulation of submarine atmosphere carbon dioxide CO_2 contamination, stabilization[1] of CO_2 content, and ventilation with fresh air.

SIMULATION TEST OF DISSUB VENTILATION

Bottles filled with $(3.50 \pm 0,01)kg$ of carbon dioxide CO_2 and $(0.25 \pm 0,01)kg$ of nitrogen N_2 were prepared for simulating preliminary CO_2 concentration into the DISSUB rescue compartment. The DISSUB rescue compartment consisted of two decks that divide the compartment into three sub-compartments of equal volume. There are stairs between decks, and a cargo hatchway[2] between the lower sub-compartments. Preliminary CO_2 concentration was created by releasing CO_2 from four bottles – two on the first deck and two on the second. There were two POLYTRON IR CO2 applied

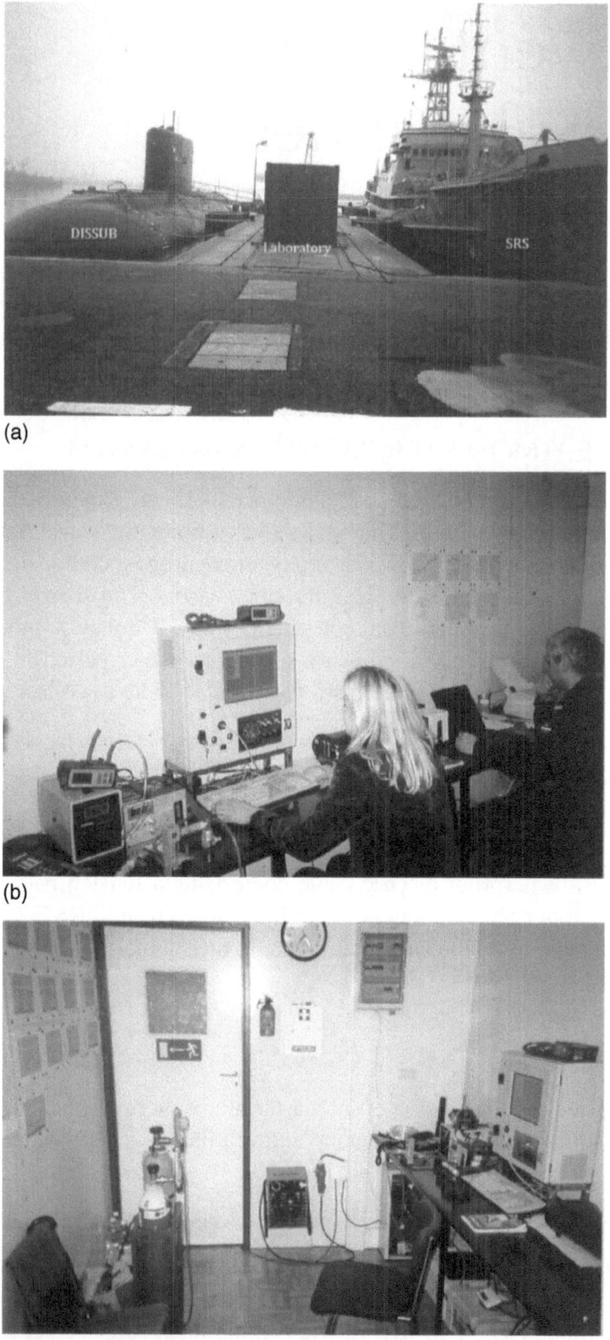

FIGURE 8.1 The portable laboratory.

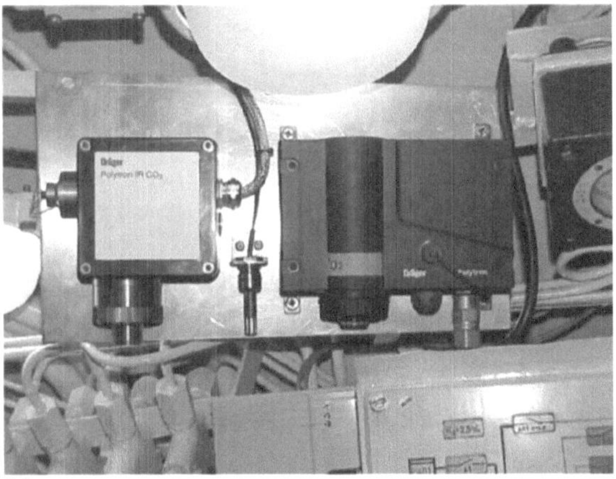

(a)

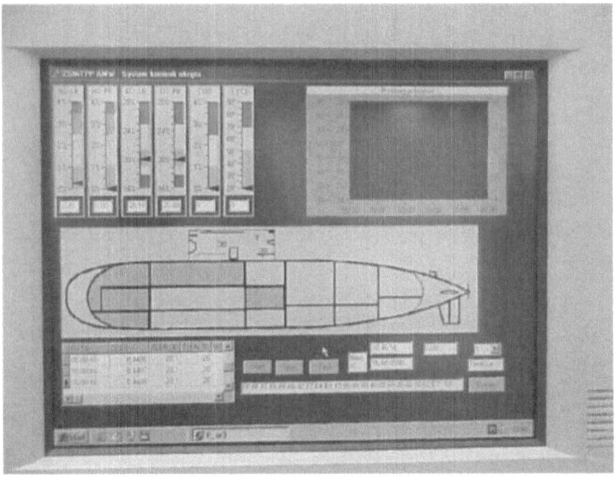

(b)

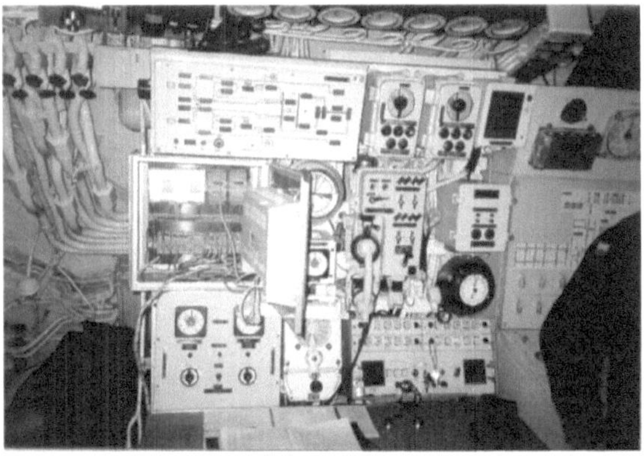

(c)

FIGURE 8.2 Oxygen and carbon dioxide gas analyzers and industrial computer as part of the experimental submarine atmospheric monitoring system: a) carbon dioxide analyzer and oxygen analyzer, b) submarine atmosphere monitoring screen, c) inside view of the PC

(Continued)

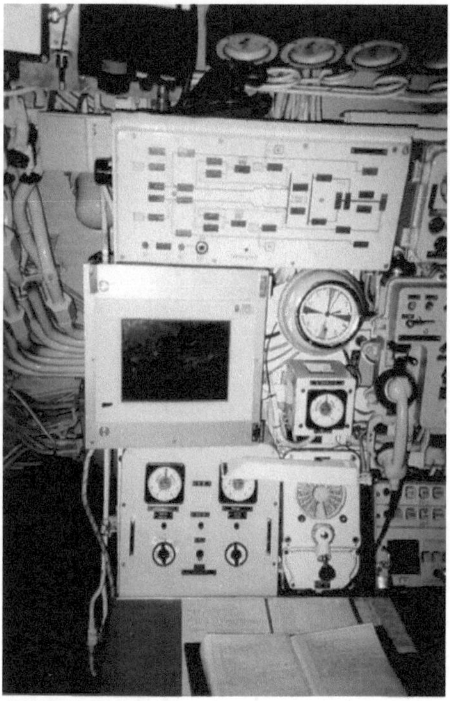

(d)

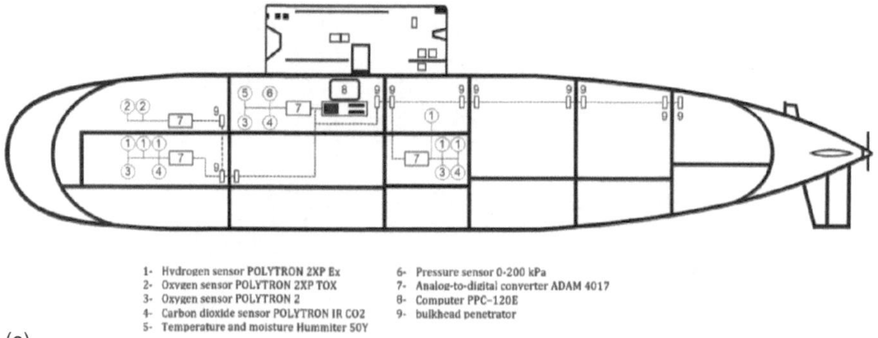

(e)

FIGURE 8.2 (Continued) d) PC box, e) scheme of the monitoring net.

for measuring CO_2 concentration by the submarine atmosphere monitoring system, on top of the upper and lower bilges of the lowermost compartment. Metrological data of POLYTRON IR CO2 were collected in Table 8.1.

HOMOGENIZATION AND VENTILATION

Homogenization of the rescue compartment atmosphere was quickly processed, taking, on average, $1h$ (Figure 8.2). The stable value of CO_2 concentration and average volume of the empty compartment could be used to calculate compartment filling by equipment: ξ.

TABLE 8.1
Main metrological data of POLYTRON IR CO2

Principle of work:	Spectrophotometer *IR*
Measurement range:	0–4.5%$_v$
Zero repeatability:	$\leq \pm 0.01\%_v$
Sensitivity:	$\leq \pm 5\%_v$
Pressure effect:	$\leq \pm 0.16\%_v$ measuring value$\cdot hPa^{-1}$
Drift:	$\leq \pm 0.4\%_v$ measuring value$\cdot$month^{-1}
Cross sensitivity:	Aldehydes, ketones, moisture

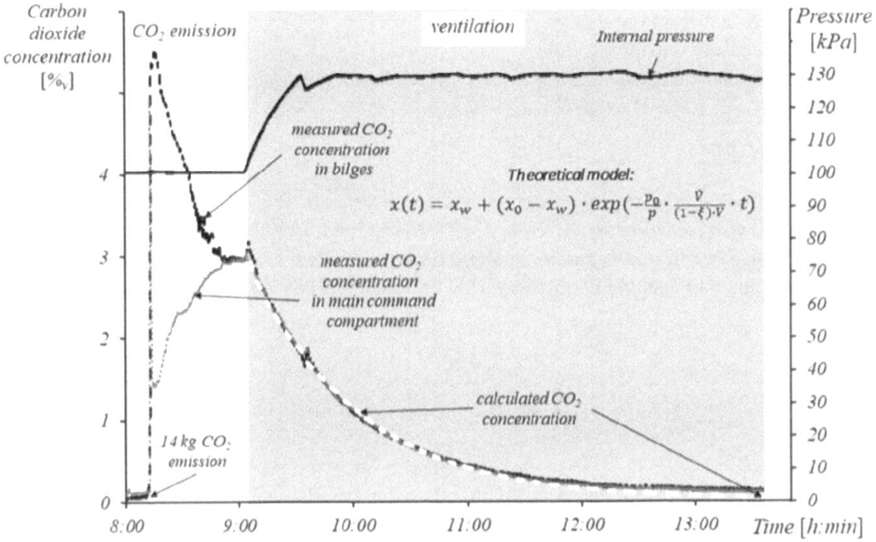

FIGURE 8.3 Results of measurements of carbon dioxide concentration in ventilated DISSUB compartment, where: average ventilation stream $\dot{V} = 4.832\,m^3\cdot min^{-1}$, carbon dioxide concentration into fresh air $x_w = 0.0002\ mol\cdot mol^{-1}$, initial concentration of carbon dioxide $x_0 = 0.0205\ mol\cdot mol^{-1}$, atmospheric pressure $p_0 = 101.325\ kPa$, rescue compartment volume $V \cong 309\ m^3$, compartment filling by equipment $\xi \cong 19.7\%$

The average stream of fresh air was calculated from pressure drop measurements in the air bank of the SRS. Derivation of the relation between carbon dioxide molar fraction as a function of time for the continuous ventilation of the habitat was shown in Tables 6.1 and 6.2. The model was indispensable to calculate the ventilation model presented in Figure 8.3 (Kłos, 2003).

Initially, it had been assumed that the pressure would rise before the ventilation process started. But ventilation started almost immediately, caused by leakage. At first a slight increase in internal pressure could be noticed, followed by a fast decrease in pressure.[3] This was caused by the DISSUB surface position – submarine seals worked in the reverse direction.[4]

SUMMARY

Good conformity between the mathematical model and the real CO_2 content measurements was recorded. Only at the end of the ventilation process a slight difference could be observed, probably caused by weak homogenization in all compartment volumes[5] for low concentration of CO_2. This exerts an influence on the effectiveness of the ventilation process. Another mathematical model is needed for modeling the ventilation process in this region, taking into consideration, e.g., diffusion resistance. But the presented model is adequate enough for practical applications (Figure 8.3).

NOTES

1 Homogenization.
2 Approximately 2 m^2 in area.
3 This is evident from Figure 8.3.
4 Higher pressure outside the hull.
5 The mathematical model did not assume perfect homogenization.

REFERENCES

Kłos R. 2003. A mathematical model of ventilation process of a distressed/disabled submarine. *Polish Maritime Research* 4, 23–25. ISSN 2083-7429.

Kłos R. 2008. *Systemy podtrzymania życia na okręcie podwodnym*. Gdynia: Polish Hyperbaric Medicine and Technology Society. ISBN 978-83-924989-4-0.

Part IV

Mining Excavations

This part describes the deterministic modeling of the mining excavation ventilation process. The dimensions of the mining excavation in which the research was carried out are comparable to that of submarine's compartments.

The excavation was located at a depth of over 900 m, hence the pressure in it was higher than the atmospheric pressure. The ventilation process, however, was different compared to that in a submarine, because the ventilation was forced by the release of gas inside the chamber, and the ventilation stream came only outside the chamber and was not directed to the surface. Hence, this process can be regarded as isobaric ventilation.

9 Ventilation of the Sealed Mining Excavation

In mining plants, there exists a hazard of occurrence of unhealthy atmosphere. Harmful pollutants can come from various sources. In coal mining, they come from the burning of coal and flammable hydrocarbons. In other mines, for example in copper mines, a fire may happen when the fuel or electrical wiring in machinery ignites, or when the gases accumulated in the rock mass seep through cracks and catch fire. During fires or flarebacks, large amounts of toxic gases can be released, which, if spread quickly, can surprise the crew, even in locations far from the fire or flareback. The smokiness of excavations occurring during a fire makes it difficult for the endangered crew to withdraw along the specific escape routes. In extreme cases, the crew will not be able to withdraw from the excavation face.

A plan for various rescue action scenarios is developed in advance to respond to cases of occurrence of hazards in a mine. The plan specifies, among other things, escape routes, which the crew may use to withdraw from the endangered zone to a safe place. In the case of long escape routes, the respiratory protection equipment used by the crew[1] may not be sufficient to ensure safe exit from the danger zone. The solution may be the construction of refuge chambers, which in the event of a fire will secure a safe haven for the crew. The possibility for the crew to survive the period of danger in such chambers gives the management of the rescue operation time necessary to take action to help them.

Due to the installation of an air-tight breathing system in the refuge chamber, when an atmosphere hazardous for safe breathing forms, a group of miners can survive for a period assumed in the chamber design.[2] The basic element of the life-saving system are sets of inhalers equipped with half-masks using air contained in high-pressure bottles. The chamber should be fitted with the necessary control and measurement equipment, communication means, etc., in accordance with the design assumptions and applicable regulations.

MINING EXCAVATION

Calculations related to the life-saving system were based on the values of the streams of oxygen used and the ventilation of lungs depending on the physical effort as shown in Table 1.13. Assuming that one air cylinder having the capacity of 50 dm^3 and working pressure of 20 MPa will be used by 2 miners, 20 air cylinders should ensure lung ventilation for 40 miners in the case of "light intensity" physical effort, according to Table 1.13.

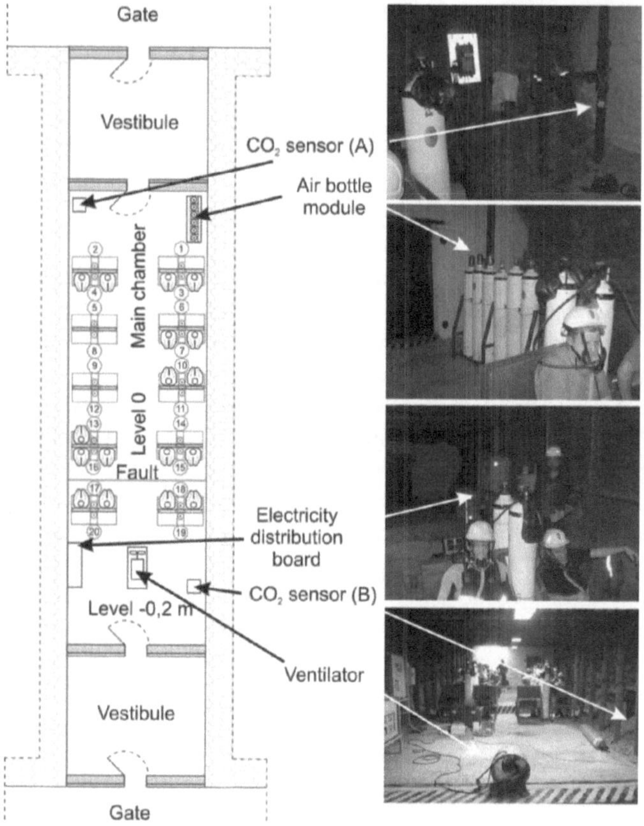

FIGURE 9.1 Arrangement of equipment items during tests.

The chamber was placed in the excavation connecting two gates having a length of min. 10 *m*, width at the thill of about 5 *m*, and height of about 2 *m*. The chamber was isolated from the gates with walled dams, in which the entrance door to the chamber was built in a sluice-box arrangement. The main dimensions of the chamber were adopted as guidelines for the construction of the life-saving system (Figures 9.1 and 9.2).

The dams were made of concrete blocks, the walls were plastered on both sides, and the ground was leveled and hardened. The tightness of the rock mass and walls sealing off the chamber from excavations was provided with generally available means.

Assuming that the chamber volume is 100 m^3 and that the emission of carbon dioxide for light-intensity effort in accordance with Table 1.13 is approx. $0.8\ c \cdot min^{-1}$, the concentration of carbon dioxide CO_2 will reach $1\%_v$ as soon as 0.5 *h* has elapsed when there are 40 people inside it. Hence, it had to be equipped with a ventilation system.[3]

The ventilation of the chamber prior to its evacuation was to be provided by one or two racks with six air cylinders of 50 dm^3 capacity and a working pressure of

(a)

(b)

FIGURE 9.2 Arrangement of equipment items during tests: a) monitoring station (in the left of the picture); b) rescuers breathing through inhalers with connected CO_2 absorbers

(Continued)

20 *MPa*, designed for ventilation of the atmosphere in the chamber just prior to the planned end of human stay inside it (Figure 9.1). The release of air was to be achieved through the total opening of the tank.

An important piece of equipment from the point of view of the tests was a fan, whose location is shown in Figure 9.1d.

(c)

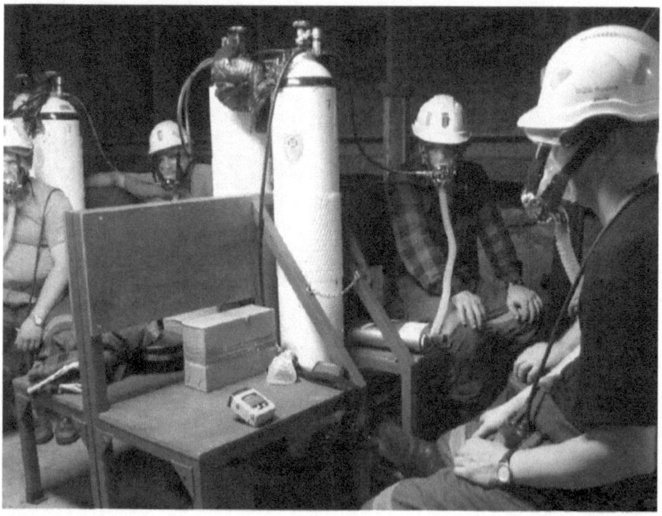

(d)

FIGURE 9.2 (Continued) c) stand for humidity and temperature; d) safety station for monitoring hazards with typical mining equipment.

MEASURING EQUIPMENT

During the tests, several measuring systems were used:

- Dräger Multi-Gas Monitor MULTIWARN II type analyzer with internal memory[4] used as a safety device to measure oxygen and carbon dioxide content (Dräger Safety AG & Co. KGaA, 2003) (Table 2.2)
- a monitoring system built by the Naval Academy, based on the RS-485 network, which used ADAM 4517 ADVANTECH analog-to-digital converter,

connected to two Dräger Transmitter Polytron IR CO2 analyzers having standard connectors[5] (Table 8.1). The results of measurements from the network were transmitted by the RS485/RS232[6] converter to a typical PC every 5 *s* (Figure 9.1a and 9.2a)

– typical portable multi-channel analyzers allowed to monitor the atmosphere in the mine are used only as additional protective devices (Figure 9.2d)

INFRARED SPECTROSCOPY

The carbon dioxide content was measured with analytical heads of the POLYTRON IR CO2, which utilize IR[7] spectroscopy. Infrared spectroscopy is a recognized and valued method of instrumental analysis, used for qualitative marking of organic compounds. The infrared spectra in the gas phase show only a slight widening of the absorption bands. This method can be used to detect trace amounts of organic compounds. Gases such as CO_2 and CO give clear and well-distinguishable absorption bands, only partially overlapping. In the range of wavenumbers from 800–2000 cm^{-1}, the spectrum is clear. This is called the range of *fingerprint* in which characteristic absorption bands of many organic compounds occur. Just as fingerprints differentiate people, bands in this range make it possible to distinguish chemical compounds as long as they can be properly interpreted. Similarly, in the range close to the wavelengths of approx. 3000 cm^{-1}, the spectrum is clear. In this range, there are hydrocarbon absorption bands. Organic compounds that may be present in the breathing atmosphere, as products of metabolism, show high absorption activity in the ranges of wavenumbers 800–2000 cm^{-1} and approx. 3000 cm^{-1}. In this way, they can be marked together with typical pollutants such as carbon dioxide and oxide.

Quantitative measurements with this method may not always be carried out with sufficient accuracy due to significant broadening of the bands, their overlapping in mixtures, inability to precisely separate and integrate them, curving of the signal dependence characteristics on the concentration,[8] and cumbersome calibration procedures. The apparatus, due to the widespread use of IR spectroscopy, is available in a form designed for measuring CO_2 content in gas mixtures.

RESEARCH INVESTIGATIONS

The MULTIWARN II type analyzer was used only as a safety device similarly to other measuring devices approved and used in mining (Table 2.2).

The basic measurement system equipped with POLYTRON IR CO2 type heads was calibrated using standard gases, guaranteeing measurements with an absolute error of $\leq \pm 0.01\%_v$. The location of CO_2 monitoring heads in the chamber is shown in Figure 9.1. They were placed at a height of about 1 *m* above ground. The MULTIWARN II type analyzer was placed at the same level near the electric panel and inhalation station N°20 (Figures 9.1 and 9.2a). Monitoring of safety carried out with typical multi-channel devices used in mining was located at stations 4, 10, and 18 (Figures 9.1 and 9.2d). In addition, periodic measurements of temperature, humidity, and air mass flows were carried out in the chamber (Figure 9.2c).

The tests were conducted for several days. Every day a test of the breathing systems was conducted with participation of people followed by a test of chamber ventilation without participation of people. One day, rescuers breathed for 1.5 h with inhalers that were not equipped with carbon dioxide absorbers. Another day, a 1.5 h breathing test was repeated using inhalers equipped with carbon dioxide absorbers (Figure 9.2d). The tests were conducted without a running fan and with the use of a fan (Figure 9.1d).

The ventilation tests were carried out without the participation of people. 8 kg of CO_2 was released into the empty and ventilated chamber, stored in two cylinders having a capacity of 50 dm^3, exactly 4 kg of CO_2 and 1 kg of nitrogen N_2 in each[9]. The cylinders were located near the sensors of the CO_2 monitoring system (Figures 9.1a and 9.1d). The addition of nitrogen was aimed at quantitatively releasing the entire contents of the cylinder. On the first day, after CO_2 concentration in the chamber was stabilized, the content of three air cylinders 20 MPa @ 50 dm^3 was sequentially released into it as quickly as possible. On the second day, the content of one 15 MPa @ 50 dm^3 air cylinder was released as slowly as possible.

RESULTS

The detailed results of the experiments with participation of people were published earlier (Kłos, 2010). They will not be discussed here in detail because they do not concern the main issue of chamber ventilation. Only general conclusions will be presented in brief:

- Contrary to theoretical predictions, the oxygen content during the tests with people practically did not change, which gives the possibility to use air to breathe, and to discontinue using breathing mixtures that generate logistic and fire-related problems.
- Despite the theoretical predictions, the CO_2 content during the 1.5 h test with people did not significantly increase,[10] which indicates that in these conditions the use of CO_2 absorbers is not required.
- The CO_2 content during the test with the fan switched on is higher than without mechanical mixing of the atmosphere in the chamber, which suggests the presence of CO_2 floating at the thill.
- The CO_2 content measured with the CO_2 sensor in the monitoring system (B) and with the MULTIWARN II during mechanical mixing of the atmosphere in the chamber with the fan did not show differences, which indicates homogenization of the atmosphere in the chamber.

The complete research results into the chamber ventilation were published earlier (Kłos, 2010) and will not be referred to here. They show that after 8 kg of CO_2 has been released into the chamber space, its concentration stabilized at $1.2\%_v CO_2$. However, sometimes there was a constant difference of $0.1\%_v CO_2$ between the opposite ends of the chamber. In the fan area, the CO_2 concentration was maintained at $1.4\%_v CO_2$ and at the opposite end at $1.3\%_v CO_2$.

DISCUSSION

The results of the final test are discussed in this section. After 8 kg *of* CO_2 was released into the chamber space,[11] the ventilation tests were carried out sequentially, based on two schemes:

- simulation of laminar flow through slow and even release of 90 m^3 of air into the chamber space for about 25 *min*
- simulation of turbulent flow of even air release of 31 m^3 to the space of the chamber for about 1 *min*

The volume of carbon dioxide V_{CO_2} released from the cylinder can be calculated as follows: $\forall_{R=const}\ V_{CO_2} = \frac{m_{CO_2}}{\rho_{CO_2}} \mid T = idem,$ where V_{CO_2} represents volume of emitted carbon dioxide, m_{CO_2} the mass of emitted carbon dioxide, and ρ_{CO_2} density of carbon dioxide and T represents temperature. And according to the general gas equation,[12] the volume of the emitted carbon dioxide $V_{CO_2}V_{CO_2}$ will be: $\forall_{R=const}\ V_{CO_2} = \frac{p}{p_0} \cdot V_k \cdot x \mid T = idem,$ where p_0 is the normal pressure, p represents chamber pressure, V_k is volume of the chamber, and x represents the molar content of carbon dioxide in the chamber. Combining these relationships, the volume of the chamber V_k can be calculated as follows: $V_k = \frac{m_{CO_2} \cdot p_0}{p \cdot x \cdot \rho_{CO_2}}$. It was assumed that the density of CO_2 in the experimental conditions was[13] $\rho_{CO_2}(t = 25°C) \cong 1.811 kg \cdot m^{-3}$. It is known from the measurements that the release of 8 kg *of* CO_2 resulted in formation of a stable concentration at the level $x \cong 0.012\ m^3 \cdot m^{-3}$ in the volume of the free chamber (Figure 9.2). Thus, the volume of the ventilated chamber was approximately $V_k \cong \frac{8\,kg}{0.012 \cdot 1.811\,kg \cdot m^{-3}} \cong 368\,m^3$.

During the tests, both with people and the ventilation tests, the results of CO_2 measurements with the CO_2 monitoring system sensor (B) and MULTIWARN II did not show any significant differences, suggesting that the distribution of CO_2 concentration was uniform across the chamber at a height of approx. 1 m. On the same basis, it can also be presumed that the measurements performed ensured sufficient accuracy, repeatability, and actuation reliability[14] during the conduct of the experiments.

Under the conditions of the experiment, no changes in the oxygen content in the atmosphere of the chamber were recorded (Figure 9.2). The ventilation model of the chamber is consistent with the mass balance of carbon dioxide in the chamber: $x_{CO_2} = x_0 \cdot \exp\left(-\frac{\dot{V}}{V_k} \cdot t\right) \ |V_k = 358\,m^3$.

In Figure 9.3, in the initial period, the stratification of the carbon dioxide content in the atmosphere of the chamber after its release from the cylinder is clearly visible.

However, after about 10 *min*, rapid homogenization and disappearance of the initial stratification took place. Such rapid homogenization could be caused by mechanical movements of the atmosphere of the chamber caused by a stream of released carbon dioxide. Initially carried out ventilation was at a lower intensity of the air

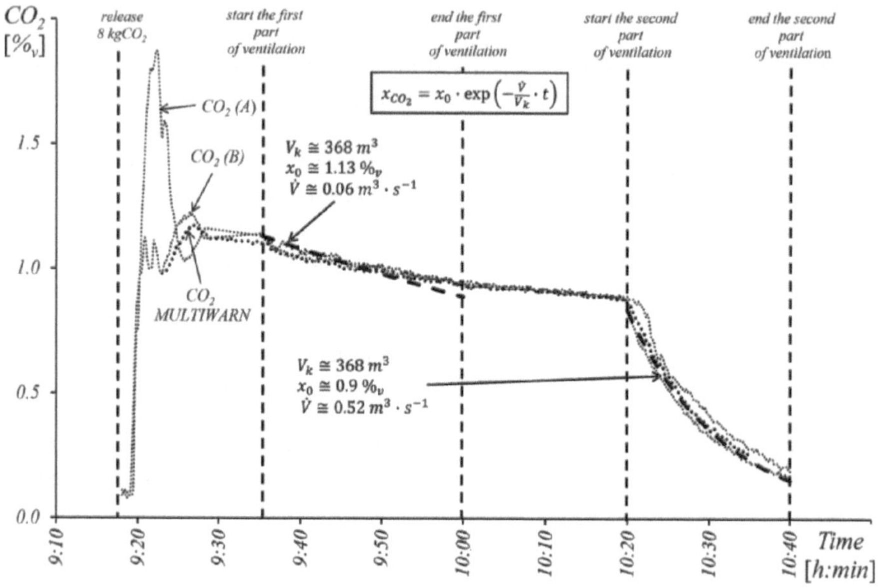

FIGURE 9.3 The results of the ventilation tests using two ventilation streams without mechanical mixing of the chamber atmosphere with a fan.

stream at the level of approx. $\dot{V} \cong 0.05\,m^3 \cdot min^{-1}$. Then, after the disappearance period of the effect of the first ventilation process, determined for approx. 20 *min*, a more intensive ventilation process was carried out using a tenfold higher air stream $\dot{V} \cong 0.53\,m^3 \cdot min^{-1}$.

SUMMARY

The described measurement results for both ventilation processes presented here demonstrate extraordinary compliance with the theoretical model. Previously, several times the results were obtained that did not show less compatibility of the theoretical model with the results of measurements.

It can be stated that the results of the research on such a large and complex object as a mining excavation are surprisingly consistent with the proposed mathematical model. This suggests that the approach used to model the ventilation process for a mining excavation is sufficiently accurate to predict its course at the assumed level of credibility, thus enabling its practical use for designing mine refuge chambers.

NOTES

1 For example, oxygen self-rescue devices and stations of spare devices.
2 Protective action time of a refuge chamber.
3 It was assumed, as for a submarine, that 1% v is an acceptable content of CO_2 in a breathing atmosphere and 2% v is the upper limit when the chamber should be vented before leaving it.

4 Data recorded every 10 s.
5 4–20 *mA*.
6 ADAM 4520.
7 *IR* – infrared.
8 Deviation from Beer-Lambert lat.
9 Samples were prepared using the weight method and had the appropriate Linde Gas certificate (Figures 9.1a and 9.1d), lying cylinders located on the right side of the drawings.
10 When carbon dioxide absorbers are not used during exhalation.
11 Without mechanical mixing of the atmosphere in the chamber with a fan.
12 *Clapeyron* equation.
13 The possibly inaccurate value of the adopted density affects the deviations from the real value of the chamber volume, but a further model of the ventilation process uses the determined value referring to the measurement conditions, hence the error caused by adoption of an inadequate value of carbon dioxide density does not affect the ventilation model.
14 The measurements showed the same changes over time.

REFERENCES

Dräger Safety AG & Co. KGaA. 2003. *Multi-Gas Monitor Multiwarn II Technical Handbook.* Lübeck: Dräger Safety AG & Co.

Kłos R. 2010. Komorowy system zachowania życia zabezpieczający górników w przypadku powstania atmosfery niezdatnej do oddychania. *Polish Hyperbaric Research* 33, 71–88.

10 Conclusions

The aim of the work was to develop and verify some mathematical ventilation models for the semi-closed-circuit rebreather[1] SCR and to generalize the presented calculation models to other hyperbaric objects, such as hyperbaric chambers, submarines, and those used in mining excavation. In most situations in a submarine and mining excavation there exists slightly higher pressure p as compared to atmospheric pressure p_a: $p \gtrsim p_a$, but for the purposes of simplification, these facilities will be classified as normobaric. The mathematical modeling presented here enables combining the micro and macro elements of the hyperbaric and normobaric technique. The developed models are of analytical character. Therefore, they have physical interpretation. Because of this, the semi-analytical and analytic models were not explored. For the sake of the analyzed phenomena character, the statistic models of phenomena were not explored either.

MANNED EXPERIMENTS WITH THE USE OF SCRs

The experiments concerning the mathematical ventilation models were performed during manned and unmanned experimental diving aimed at verification of the decompression procedure proposed. The experiments were carried out for SCRs. However, the tested SCRs should be characterized according to the same parameters as those assumed at derivation of the mathematical model. It is the main condition necessary to perform the experiments.

Before verifying the mathematical model of the SCR with constant breathing gas dosage by means of a metabolic simulator, a series of experimental dives were carried out. To test the mathematical model at the upper and lower limits of oxygen partial pressures in the inhaled breathing gas, two different methods were applied. The lower limits were verified during the basin tests with the use of SCR APW-6 M SCUBA diving apparatus (Chapter 3). The adequacy of the assumed mathematical model at upper limits of oxygen partial pressures was tested during the pressure tests with the use of the diving apparatuses type SCR GAN-87 UBA. The experiments were carried out in parallel with verification experiments of the decompression procedure. There were more similar trials carried out, but the system was modified.[2] That is why the results obtained are not sufficient to draw the conclusions concerning the adequacy of the mathematical ventilation model. The probability of the proposed mathematical model acceptance[3] was not further maximized because of the high experimental costs.

It seems that the best method for experimentation is to apply a breathing machine simulating breathing action with simultaneous oxygen sampling from the breathing space and carbon dioxide emission for unmanned tests. Carbon dioxide emission is maintained at the proper level,[4] in which case there is no necessity to carry out manned experiments and the experimental results are repeatable with known accuracy of reproduction. Within the framework of the financially supported project, a special laboratory set was built. It was to enable the performance of less expensive experiments and to get the results without the necessity to conduct manned experiments. The mathematical model was verified during the unmanned tests with the use of SCR CRABE SCUBA diving apparatus (Chapter 4).

The preliminary operating assumptions for design of the SCR with the constant breathing medium dosage are given in Table 1.14. The table was prepared on the basis of multiple solution of the equation set (1.10). The preliminary exploitation assumptions for the SCRs types APW-6M, GAN-87, and FGG III[5] are given in Table 1.14. For the SCR of the type APW-6M it was assumed that the fresh breathing medium metering is maintained at the level of 8 $dm^3 \cdot min^{-1}$ for two operational gas mixtures $Nx\ 55\%_v O_2/N_2$ and $Nx\ 45\%_v O_2/N_2$ (Table 1.4). Verification of the assumptions is presented in Figures 2.14 and 3.3. It was carried during distance swimming at shallow depth. According to the medical and technical requirements for those trials, the experiments carried out confirmed that assumption. That stage of experiments was the introduction to the experimental diving at the depth of 20 mH_2O. The pressure tests have confirmed the above-mentioned assumptions, despite the fact that selected metering rates were lower than those obtained from calculations. The oxygen partial pressure drops observed in the breathing space were higher than intentional ones. Therefore, it was compensated by decompression time increase. It was particularly observed during diving where the divers performed intensive exercise within the depth range of $[20;30] mH_2O$. The divers breathed with the breathing mixture of the composition $Nx\ 45\%_v O_2/N_2$. The observed oxygen partial pressure drop was consistent with the proposed ventilation mathematical model. The choice of the lower fresh breathing medium rate gave the possibility of extending the apparatus protective functioning time to the same amount of breathing medium from an integral unit of supply flasks. Experimentally verified guaranteed protective functioning time of the carbon dioxide scrubber is 3 h at the water temperature not less than 5°C.

The preliminary operating data for training use[6] of the SCR type GAN-87 was presented in Tables 1.6–1.9. There are the different working modes of the multi-nozzle SCR type GAN-87. The working mode of the apparatus is depth dependent. According to the tactical and technical assumptions, the training version of the apparatus should enable diving to the average depths.[7]

Investigations of the diving apparatus were performed at the following parameters: oxygen contents $Nx\ 27\%_v O_2/N_2$ and gas metering 25 $dm^3 \cdot min^{-1}$. Maintaining oxygen partial pressure in the inhaled breathing gas within the limits of 20–160 kPa is the necessary condition for an experimental confirmation of the previously made assumptions. The results obtained during diving at the greater depths[8] aimed at verification of the upper limit of the oxygen permissible partial pressure in the inhaled breathing medium. It is evident from Figure 2.14 that the limit of 160 kPa is not exceeded. For the depth of 50 mH_2O, it is shown that the limit was reached[9] when the diver was at rest. Exceeding the depth limit 50 mH_2O even by 5 mH_2O causes evident exceeding of the limit of 160 kPa. The experiments performed have shown that assumptions concerning maximum oxygen partial pressure in the inhaled breathing medium have been confirmed.

The mathematical ventilation model of the SCR with constant gas metering was verified using statistical inference, which confirmed the hypothesis of sufficient adequacy of the proposed mathematical model at the confidence level of 95%.

UNMANNED EXPERIMENTS WITH THE USE OF SCRs

In order to simulate the breathing process, it is necessary to solve two technical problems: mechanical representation of pulmonary ventilation and simulation of the gas exchange during breathing. The processes should be simulated with sufficient accuracy and repeatability. There is no need for representation of each detail of the processes. It is not necessary to simulate the real ventilation process. Usually it is assumed that the respiration shape should be approximated to sinusoidal character. As a matter of fact, each of the hyperbaric research centers has their own evaluation standards. A series of standards concerning the respiration volume and the respiration rate have been established.

There are many similar regulations in the world literature. For example:

- According to Norwegian Underwater Technical Centre NUTEC, the maximum total breathing work (Segedal & Morrison, 1985) can be expressed by empirical formula $w = 0.5 + 0.02 \cdot \dot{V}$, where w represents normalized work of breathing $[w] = J \cdot dcm^{-3}$ and $\dot{V}$ lung ventilation $\left[\dot{V}\right] = dm^3 \cdot min^{-1}$. Apart from that, the maximum positive gauge pressure[10] in a mouthpiece assembly should not exceed 2.5 *kPa*.
- Det Norske Veritas DnV set up a regulation that for ventilation of the level 62.5 $dm^3 \cdot min^{-1}$ at 25 min^{-1} breathing cycles, the mean breathing work should not exceed $\dot{w} = 1.7 J \cdot dm^3 \cdot min^{-1}$.
- US Navy Experimental Diving Unit NEDU (Segedal & Morrison, 1985; Reimers & Hansen, 1972) set up a regulation that the maximum positive/negative gauge pressure in the mouthpiece assembly for the corresponding ventilation should not exceed values given in Table 10.1. The values of breathing work[11] are calculated according to the sinusoidal approximation.
- In the European Union, the standard EN 250 Respiratory equipment – Open Circuit Self Contained Compressed Air Diving Apparatus – Requirements, Testing and Marking defines minimum performance standards (PN-EN 250, 2014).

TABLE 10.1
Maximum positive/negative gauge pressure in the mouthpiece assembly and maximum permitted breathing work (NEDU, 1994)

Pulmonary ventilation	Maximum breathing resistance	Maximum normalized breathing work
[$dm^3 \cdot min^{-1}$]	[*kPa*]	[$J \cdot dcm^{-3}$]
22.5	0.10	0.13
40.0	0.45	0.60
62.5	1.00	1.33

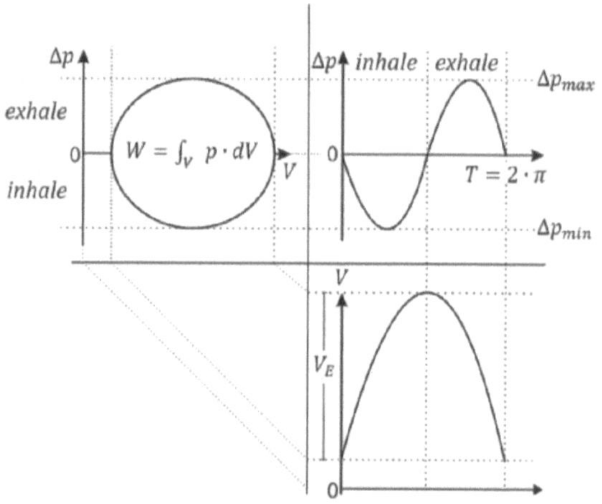

FIGURE 10.1 The changes in pressure and the breathing space volume during respiration, sinusoidal approximation, where: $\dot{V}$ – pulmonary ventilation, W – breathing work, $\Delta p(t)$ – the pressure changes in the mouthpiece as a function of time t, t – time, T – period, V – breathing capacity.

The sinusoidal approximation is one of the methods applied to evaluate breathing work. The cyclic changes in pressure and breathing work are presented in Figure 10.1. The breathing work can be determined according to the sinusoidal approximation:

$$W = -\int_t \Delta p(t) \cdot \frac{\partial V(t)}{\partial t} \cdot dt = -\int_t \Delta p(t) \cdot \dot{V}(t) \cdot dt \tag{10.1}$$

where $\dot{V}$ is pulmonary ventilation, W is breathing work, $\Delta p(t)$ is the pressure changes in the mouthpiece as a function of time t, $\frac{\partial V(t)}{\partial t}$ is the breathing capacity changes as a function of time, t is time, and T is period.

Following the sinusoidal approximation, the pressure changes in the mouthpiece as a function of time can be written as follows:

$$\Delta p(t) = -\Delta p_{max} \cdot \sin(\omega \cdot t) \tag{10.2}$$

where Δp_{max} is maximum/minimum pressure in the mouthpiece during breathing action, pulsatance (angular frequency: $\omega = 2 \cdot \pi \cdot f = 2 \cdot \pi \cdot \frac{1}{T}$), and f is the breathing frequency.

The breathing capacity changes as a function of time can be written as follows: $V(t) = -V \cdot \sin(\omega \cdot t)$, thus: $\frac{\partial V(t)}{\partial t} = -\omega \cdot V \cdot \sin\left(\frac{\omega}{2} \cdot t\right) \cdot \cos\left(\frac{\omega}{2} \cdot t\right)$ and $\frac{\partial V(t)}{\partial t} = \frac{\omega}{2} \cdot V \cdot \sin(\omega \cdot t)$.

In the function $\dot{V} = f(t)$ the period is $T = 2 \cdot \pi$, hence the function amplitude is $\frac{1}{2} \cdot V : \dot{V} = \frac{1}{2} \cdot V \cdot \sin(\omega \cdot t)$. the breathing work is given as:

$$\begin{aligned} W &= \frac{1}{2} \cdot \Delta p_{max} \cdot V \cdot \int_{t=0}^{t=T} \sin^2(\omega \cdot t) \cdot dt = \\ &= \frac{1}{2} \cdot \Delta p_{max} \cdot V \cdot \left[\frac{\omega \cdot t}{2} - \frac{\sin(2 \cdot \omega \cdot t)}{4} \right]_0^{\frac{2 \cdot \pi}{\omega}} = \\ &= \frac{\pi}{2} \cdot \Delta p_{max} \cdot V \end{aligned} \tag{10.3}$$

The breathing work determined is standardized, i.e., recalculated per 1 dm^3 of the respiratory volume at a given depth of diving. The breathing work determined with the above method can be expressed as $w(\dot{V}, H) = \frac{\pi}{2} \cdot \Delta p_{max} \cdot V$, where w represents normalized breathing effort and H the depth of diving. The breathing work determined from the above formula $w(\dot{V}, H) = \frac{\pi}{2} \cdot \Delta p_{max} \cdot V$ is called resistive effort (Reimers & Hansen, 1972). The absolute values of resistive effort expressed in those units are equal. In order to calculate the breathing work using the sinusoidal approximation at the constant ventilation, it is sufficient to measure only the maximum value of pressure Δp_{max} during inhalation and exhalation. Therefore, the inhalation and exhalation work should be determined. Summarization of the work calculated gives the total work. In order to establish the same measurement conditions it was stated that resistive effort should be determined for the typical pulmonary ventilation dependent on the work performed. For those purposes, the model of the standard man was developed. It is presented in Table 10.2 (Segedal & Morrison, 1985).

Investigating the problems concerned with ventilation, other additional parameters are studied so that they can be simulated.[12] The above-mentioned problem has not been solved yet (PN-EN 250, 2014). For these and other reasons it was assumed

TABLE 10.2
The relationship between the pulmonary ventilation, inhalation volume, and the stream of emitted carbon dioxide and kind of effort for the standard man

Pulmonary ventilation	The stream of emitted CO_2	Breathing frequency	Inhalation volume	Kind of work performed
$[dm^3 \cdot min^{-1}]$		$[s^{-1}]$	$[dm^3]$	
15.0	0.6	15	1.0	Rest
22.5	0.9	15	1.5	Light
40.0	1.6	20	2.0	Medium
62.5	2.5	25	2.5	Moderately hard
75.0	3.0	25	3.0	Hard
90.0	3.6	30	3.0	Extremely hard

that for the mechanical simulation of the pulmonary ventilation the piston machines are applied. However, other constructions are applied as well (Minilab Mod. No. 92-25, 1992). At present, the Diving Gear and Underwater Work Technology Department is equipped with the piston-type breathing machine.

During simulation of the breathing action and gas exchange, the exact representation of the inhalation and exhalation shape is not essential. Attention should be focused on the oxygen consumption and carbon dioxide emission to that place. The two most often applied values of the *respiratory quotient* $\varepsilon_{RQ} = \{0.8; 0.75\}_v$, which shows the volume ratio $\varepsilon_{RQ} = \frac{\dot{v}'}{\dot{v}}$, describe the ratio of carbon dioxide output $\dot{v}'$ to oxygen intake $\dot{v}$. When carbohydrates are metabolized in the human body, it is determined at the level $\varepsilon_{RQ} = 1.0\ dm^3 \cdot dm^{-3}$; for proteins, $\varepsilon_{RQ} \cong 0.8\ dm^3 \cdot dm^{-3}$; and for fats, $\varepsilon_{RQ} \cong 0.7\ dm^3 \cdot dm^{-3}$. It is possible to use a reactor with a catalyst for burning chemical compounds as a metabolic simulator. The combusted chemical substance is added to the drawn volume from the breathing loop of the UBA using a breathing simulator through the inhale hose. After passing through the reactor, this mixture is returned together with the combustion products to the breathing loop of the diving apparatus through the exhale hose. It is possible to realize it according to the following reactions (2.1):

$$2 \cdot CH_3CHO + 5 \cdot O_2 \xrightarrow{catalyst} 4 \cdot CO_2 + 4 \cdot H_2O \quad \varepsilon_{RQ} = 0.80$$

$$(CH_3)_2 CO + 4 \cdot O_2 \xrightarrow{catalyst} 3 \cdot CO_2 + 3 \cdot H_2O \quad \varepsilon_{RQ} = 0.75$$

In order to dose acetic aldehyde or acetone, it is necessary to use an accurate pump, supplying the given liquid stream at a pressure not exceeding 1 *MPa*. The dosage of acetic aldehyde or acetone should be maintained at the approximate level of 0.50–4.00 ± 0.05 $cm^3 \cdot min^{-1}$, assuming that the diver's oxygen consumption is within range be $\dot{v} \in 0.3 - 2.4\, cm^3 \cdot min^{-1}$ (Table 1.13). As the catalyst, the most popular platinum-based catalyst will be used. The research concerning the above problems is carried out by the author within the framework of the grant awarded (see Chapters 2 and 5 for details).

At the first approach, the required rate of acetone was maintained by means of the pump type 40-PS4L, manufactured by Lubrizol Company. The pump characteristic was checked. The satisfactory accuracy of counterpressure liquid metering to 1 *MPa* was obtained. 0.5% platinum deposited on rollers of alumina was used to research acetone oxidation. Combustion products passed through gas chromatograph metering valve coupled with mass spectrometer GC/MS or gas analyzers set type Multiwarn II. The composition of gas mixture that leaves the top part of the catalyst bed was measured at different reaction temperatures on the border of the reaction center –350–450°*C*. There were no acetone traces in spectra GC/MS recorded. Sometimes not significant 100–200 ppm_v carbon monoxide contents were noticed during the experiments. Increase in carbon monoxide contents was due to the increase of the catalyst temperature in the reaction zone. It may be caused by incomplete combustion when percentage contents of oxygen drops below 10% by volume. *Factorial analysis* of the preliminary results shows that temperature exerts a large effect on the reaction kinetics and therefore on carbon monoxide emission. As a matter of fact, temperature exerts effect on the catalyst effectiveness. It seems that intensive, internal parallel current cooling of the whole reactor would be successful. At the border of the reaction core the reaction temperature should be maintained at the level of approximately of 200°*C*. As it follows from the additional experiments, at temperatures less than 80°*C* the course of reaction is weak and below 50°*C* the reaction decays. Additionally, maintaining the temperature below 100°*C* causes the water condensation in the outlet of the reactor, which leads to some difficulties. To initiate the reaction and thermal stabilization inside, the reactor casing should be initially and periodically heated, outside and inside. The reactor was filled up with the catalyst to the approximate height of 15 *cm* and diameter 70 *mm*. The cooler was immersed in the catalyst layer above the dosing pipe ending in a nozzle.

To test the reactor temperature, different test runs were performed. As the tests' result, 290 measuring points were obtained. The temperatures were recorded in a computer system and CO_2, CO, and O_2 contents were recorded in the memory of the *Multiwarn* gas analyzer. It follows from the analysis of the mean values obtained in the individual measurements that the real repeatability of the measuring stand in these conditions is $\pm 0{,}2\%_v CO_2$. This value is located on the border of the real measurement accuracy of the gas analysis methods being applied. A good agreement of the experimental results was obtained with the theoretical models in the case of water vapor in reaction products: $x_{O_2} \cong 0.208 - 1.436 \cdot x_{CO_2}$, with correlation coefficient >0.99 (Table 2.13).

A tested attachment that enables oxygen uptake and carbon dioxide emission was connected to the breathing machine. The experimental results concerning the breathing gas composition were recorded with *Multiwarn* gas analyzer. In addition to oxygen content, *Multiwarn* carried out CH_4 equivalent to organic compounds, CO and CO_2 contents. CH_4 and CO_2 contents were below measuring threshold, whereas CO contents reached up to 100 ppm_v. In the case of oxygen contents, a decrease below approx. $10\%_v$ causes some problems with reaction stabilization,[13] as the combustion process on the catalyst gets weaker or comes to a stop. It is a typical phenomenon known from the fire law of thermodynamics.

That simulator was developed and used to verify the mathematical model of ventilation developed for SCRs with the constant *premix* metering and construction with a metering bellows dispenser. Owing to the experiments carried out, it was possible to test the exactness of the theoretical ventilation model in SCRs. Use of the metabolic simulator enabled verification of this thesis. Verifying that thesis was possible with the use of a metabolic simulator.

The developed experimental laboratory stand enables continuing research on ventilation in various UBAs. Further research focused on developing reactors that could be used for the other research purposes such as regeneration of the ecologically confined space, e.g., during saturation dives or in submarines, is recommended.

EXPERIMENTS ON THE SUBMARINE AND SEALED MINING EXCAVATION

The scientific measures described here were aimed at finding out whether the mathematical ventilation models applied for the SCRs, hyperbaric chambers, sealed mining excavation, and submarines ventilation mathematical model.

There is good conformity between the mathematical model and the real CO_2 content measurements from experiments. Only at the end of the ventilation process a slight deviation was observed, probably caused by weak homogenization into all compartment volumes.[14] This exerts an effect on the effectiveness of the ventilation process. Another mathematical model is needed for modeling the ventilation process in this region, taking into consideration factors such as diffusion resistance. The presented model of DISSUB ventilation is enough for practical applications. Sufficient correspondence between the proposed mathematical model of DISSUB ventilation and real data gives a chance to predict with sufficient accuracy the ventilation parameters needed during submarine rescue operations.

SUMMARY

The basic issue in designing objects intended for people to stay inside them is to provide these objects with the right amount and quality of breathing gas. The ventilation process of systems with closed or semi-closed circulation of breathing gas, in particular hyperbaric facilities,[15] causes many additional medical and technical problems that do not occur inside typical objects with normobaric pressure.[16]

In an atmosphere that fills an ecologically closed or semi-enclosed system, there is always some accumulation of pollutants and a need to replenish the used oxygen. In hyperbaric conditions, acceptable concentrations of many pollutants are usually smaller by an order of magnitude (Kłos, 2017). Also, the oxygen content must be kept within narrower limits because of its carcinogenic and toxic properties (Kłos, 2014a; Kłos, 2014b; Kłos, 2014c; Kłos, 2014d). Such action of oxygen and pollution in hyperbaric conditions constitutes a physiological problem that is difficult to solve (Kłos, 2012).

To control concentration of pollutants and oxygen in conditions of increased pressure, greater measurement accuracy than in normobaric conditions must be ensured. This makes it necessary to use laboratory-class devices in the conditions typical for these systems to operate (Kłos, 2015b). In practice, it is difficult to use – for example, mass spectrometry methods in a diving bell – because there is a lack of analyzers of this type, which are both more affordable and resistant to severe operating conditions (Kłos, 2000). It is for this reason that research investigations that allow increasing accuracy of measurement methods are important in ensuring safety for people who stay in this environment for a long time. They include chemometric studies[17] using data mining methods.[18]

As already mentioned, the parameters of respiratory environment are the most difficult to provide inside objects that are ecologically closed or semi-closed systems. These include hyperbaric complexes, submarines, bunkers, survival chambers, etc. The results of the research aimed at improving safety conditions became apparent after the tragedy of the *Kursk* submarine. Rescuing the crew of a submarine or divers is currently a global problem, and the results of research on the safety of such operations are publicly available, unlike research on ventilation processes in other military objects, such as shelters. NATO specialists are required to disseminate knowledge in the field of safety, which includes problems related to ventilation processes.[19] The results obtained are general solutions and can be applied to typical buildings.[20]

The search for mathematical models is a commonly used approach. When dealing with complex issues, various methods are used to search for empirical or semi-empirical models. With regard to less complicated problems, attempts are made to develop analytical models that have physical interpretation. Validating the latter is often unnecessary because they are based on the verified laws of nature. The author has conducted research aimed at development and verification of a general analytical mathematical model of ventilation process, both for objects with semi-closed and closed circulation of breathing gas, for hyperbaric and normobaric conditions.

This work attempts to verify the general mathematical model developed by the author of the hyper- and normobaric ventilation process by conducting research on selected, possibly diversified real objects. For the research, a diving apparatus with closed *premix* circulation was selected as a model of a small hyperbaric object, and a hyperbaric chamber was selected as a model of a large object. A submarine was selected as a model of a large normobaric object with closed circulation of breathing gas, and a mining excavation was selected as a model of a semi-closed object.

The research was carried out over a period of about 30 years and initially was not aimed at finding and verifying the general, analytical mathematical model of the ventilation process. The cycles of research resulted from the needs of the Armed Forces of the Republic of Poland, and above all the Polish Navy. In the course of the research, however, a concept to generalize the applied approach was developed, which was used by the author to develop an adequate, with regard to practical applications, analytical mathematical model of the ventilation process in such objects (Kłos, 2000; Kłos, 2007a).

The presented approach can be extended to other military objects[21] and can be used in civilian[22] applications, especially when faced with a terrorist threat. This approach can be directly employed to the design of ventilation systems for chemical industry factories or air-conditioned buildings. Due to the specific properties of such buildings, efforts are made to seal them completely.

The account of the research relates to the types of object, and the tests are described in a chronological order. At the beginning, the modeling of the ventilation process in diving apparatuses with semi-closed circulation of *premix* and its validation tests are discussed (Kłos, 2000; Kłos, 2012; Kłos, 2016). Here, to maintain the order in which the types of object are described, included are most recent studies, also being conducted now, which have not been published yet. Next, the author generalizes the presented approach to hyperbaric chambers and submarines (Kłos, 2000; Kłos, 2008; Kłos, 2010). Designing and modeling regenerative breathing apparatuses and assessment of their operating safety with regard to the assumed diving technology are based primarily on learning and predicting the changes occurring during the ventilation process in their breathing space. The verified mathematical model of the ventilation process in diving apparatuses facilitates computer-supported simulation analyses. In the 1990s, a research project was carried out at the Naval Academy to develop a national diving methodology to use of semi-closed rebreathers with constant dosing of *premix*. As a result of the work done and the analysis of diving accidents, it was assumed that the mathematical models used at that time were not sufficiently accurate to develop optimal diving procedures.

The macroscopic models of ventilation in a semi-closed rebreather with constant dosing of *premix* proposed by the author required precise verification. Such a procedure is necessary to secure survival of people in extremely unfavorable conditions. Conduct of this kind of work for the needs of aviation, as well as in the space industry and underwater activities, is governed by many strict regulations. The final trial always embraces experiments with participation of people. To carry out these experiments, a permission by the Ethics Committee for Scientific Research should be obtained. The studies conducted by the author based on simple statistical inference from the results of the experiments with people confirmed the sufficient accuracy of the analytical mathematical model used for decades (Kłos, 2000; Kłos, 2002b). A critical analysis of this approach, however, showed some deficiencies in elaborating the research results with this method. The author, therefore, proposed a research method based on the use of a simulator. The original position of the metabolic simulator of gas exchange developed by the author guaranteed reproducible measurement conditions (Kłos, 2002a; Kłos, 2019). The results of our own research showed that the modeling method used so far was not a sufficiently accurate approximation of the processes taking place. As use of the previous model might lead to deviations from the assumed working conditions, which were potentially dangerous for the diver, and difficult to detect when the existing research procedures were applied, the author proposed to modify it (Kłos, 2002b).

CONSUMMATION

The result of the research carried out and described here is the development and verification of the approach to modeling of the ventilation processes in semi-closed and closed breathing gas circulation operated in normobaric conditions and under increased pressure.

As shown, models of the ventilation process for semi-closed rebreathers with constant dosing of *premix* used for years were not incorrect but less accurate than the ones presented and verified experimentally by the author. The research on the process of ventilation of the hyperbaric complex in a submarine and mining survival chamber confirmed the effectiveness of the proposed approach to modeling of the ventilation process initially used to model the ventilation of diving apparatus, which allowed making this approach generally applicable as a methodology (Kłos, 2000; Kłos, 2002a; Kłos, 2003a; Kłos, 2003b; Kłos, 2007b; Kłos, 2019). The conducted research led to development procedures used in diving and in ventilation in submarines and hyperbaric complexes. Their review and modification were forced by diving accidents and work on new technologies. Therefore, the presented sequence of actions was not caused by a specific research idea, but rather a response to the demand. Some research on the issues of regenerative ventilation not described here was carried in large hyperbaric facilities (Kłos, 2008; Kłos, 2009; Kłos, 2015a).

Efforts to seal industrial objects in order to protect people working there against external pollution have been taken for a long time. In typical residential and public buildings, similar methods have been used relatively recently. In general, they involve air-conditioning of the sealed buildings. This entails the need to regenerate the respiratory atmosphere inside, making them similar to military facilities. Also, implementing civil defense recommendations typical buildings have to meet the requirements concerning protection of people. Methods of protection against contamination can be used with regard to atmospheric pollution, especially in the work environment. As in the case of military facilities, the methods of modeling ventilation in standard and hyperbaric objects described here would allow developing more accurate methods to design and use ventilation and air-conditioning systems in buildings. Using them in some buildings is justified in a period of increased terrorist threat.

NOTES

1 With constant-dosage breathing medium and construction with a metering bellows dispenser.
2 The parameters assumed at derivation of the mathematical model.
3 Further calculated.
4 Metabolic breathing simulator.
5 After modification as construction similar to GAN-87.
6 Diving with the use of nitrogen-oxygen gas mixture *Nx*.
7 Up to 50 mH_2O.
8 Diving at the smaller depths is omitted.
9 There is even tendency for its exceeding.
10 Negative gauge pressure.
11 Related to resistance.
12 for example the elasticity
13 Meaning, at sufficiently large acetone dosage.
14 The mathematical model was assumed to have perfect homogenization.
15 For example, caisson, hyperbaric chamber, diving bell, etc.
16 For example, in offices, conference rooms, lecture halls, special military facilities, etc.
17 Chemometrics is a new field of knowledge whose name was developed analogically to econometrics and expresses the search for statistical models of phenomena, following the paradigm in physics that refers to inherence of stochastic phenomena existing in deterministic systems; ways to increase accuracy of measurements by means of their more advanced metrological service, as in the *SixSigma* system, are interesting.
18 Data mining, data digging methods.
19 Since 1998, the author has been involved in the work of the Underwater Working Group NATO Standardization Organization.
20 Such as office rooms, conference rooms, lecture halls, shelters, etc.
21 For example, shelters, sealed combat vehicles, sealed surface vessel, etc.
22 Air-conditioned living and office space, public utility rooms, etc.

REFERENCES

Kłos R. 2000. *Aparaty Nurkowe z regeneracją czynnika oddechowego*. Poznań: COOPgraf. ISBN 83-909187-2-2.

Kłos R. 2002a. Metabolic simulator supports diving apparatus researches. *Sea Technology* 12, 53–56.

Kłos R. 2002b. Mathematical modelling of the breathing space ventilation for semi-closed circuit diving apparatus. *Biocybernetics and Biomedical Engineering* 22, 79–94.

Kłos R. 2003a. A mathematical model of ventilation process of a distressed/disabled submarine. *Polish Maritime Research* 4, 23–25.

Kłos R. 2003b. Experimental verification of a new mathematical model of ventilation of closed circuit breathing apparatus. *Polish Maritime Research* 1, 25–30.

Kłos R. 2007a. Modelowanie procesów wentylancji obiektów normo- i hiperbarycznych. Higher PhD diss., The Polish Naval Academy. PL ISSN 0860-889X nr 160A.

Kłos R. 2007b. *Mathematical modelling of the normobaric and hyperbaric facilities ventilation*. Gdynia: Polish Hyperbaric Medicine and Technology Society. ISBN 978-83-924989-0-2.

Kłos R. 2008. *Systemy podtrzymania życia na okręcie podwodnym*. Gdynia: Polish Hyperbaric Medicine and Technology Society. ISBN 978-83-924989-4-0.

Kłos R. 2009. *Wapno sodowane w zastosowaniach wojskowych*. Gdynia: Polish Hyperbaric Medicine and Technology Society. ISBN 978-83-92499889-5-7.

Kłos R. 2010. Komorowy system zachowania życia zabezpieczający górników w przypadku powstania atmosfery niezdatnej do oddychania. *Polish Hyperbaric Research* 33, 71–88.

Kłos R. 2012. *Możliwości doboru ekspozycji tlenowo-nitroksowych dla aparatu nurkowego typu AMPHORA – założenia do nurkowań standardowych i eksperymentalnych*. Gdynia: Polish Hyperbaric Medicine and Technology Society. ISBN 978-83-924989-8-8.

Kłos R. 2014a. Inherent unsaturation. The risk of central nervous system oxygen toxicity part 1. *Polish Hyperbaric Research* 1, 37–64. DOI: HTTP://DX.DOI.ORG/10.13006/PHR.

Kłos R. 2014b. The pathophysiology related to the toxic effect of oxygen. The hazard of central oxygen toxicity part 2. *Polish Hyperbaric Research* 2, 15–34. DOI: HTTP://DX.DOI.ORG/10.13006/PHR.47.2.

Kłos R. 2014c. Survival analysis. The risk of central oxygen toxicity part 3. *Polish Hyperbaric Research* 3, 33–48. Tom 48, DOI: HTTP://DX.DOI.ORG/10.13006/PHR.48.3.

Kłos R. 2014d. The hazard of central oxygen toxicity occurrence. The risk of central oxygen toxicity part 4. *Polish Hyperbaric Research. 4*, 19–31. DOI: HTTP://DX.DOI.ORG/10.13006/PHR.49.2.

Kłos R. 2015a. *Katalityczne utlenianie wodoru na okręcie podwodnym*. Gdynia: Polish Hyperbaric Medicine and Technology Society. ISBN 978-83-938-322-3-1.

Kłos R. 2015b. Measurement system reliability assessment. *Polish Hyperbaric Research.* 2, 31–46. DOI: 10.1515/phr-2015-0009.

Kłos R. 2016. *System trymiksowej dekompresji dla aparatu nurkowego typu CRABE*. Gdynia: Polish Hyperbaric Medicine and Technology Society. ISBN 978-83-938322-5-5.

Kłos R. 2017. Pollutions of the hyperbaric breathing atmosphere. *Scientific Journal of Polish Naval Academy.* 208, 31–44. DOI: 10.5604/0860889X.1237621.

Kłos R. 2019. Modelling of the Nnormobaric and Hyperbaric Facilities Ventilation. *International Journal of Mechanical Engineering and Applications.* 1, 26–33. DOI: 10.11648/j.ijmea.20190701.14.

Minilab Mod. No. 92-25. 1992. *Operators manual*. Husøysund: Ottestad Breathing Systems.

NEDU. 1994. *U.S. Navy unmanned test methods and performance goals for underwater breathing apparatus*. Panama City: US Navy Experimental Diving Unit. Report No 01-94.

PN-EN 250. 2014. *Respiratory equipment – Open-circuit self-contained compressed air diving apparatus – Requirements, testing and marking*. Brussels: CEN-CENELEC Management Centre.

Reimers S.D. & Hansen O.R. 1972. *Environmental control for hiperbaric applications*. Panama City: US Navy Experimental Diving Unit. NEDU Rep. 25–72.

Segedal K. and Morrison J.B. 1985. *Acceptable criteria and unmanned test procedures for underwater breathing apparatus*. Bergen: Norwegian Underwater Technical Centre, 1985. Report NUTEC 17–85.

References

Kłos R. 1990. Metodyka pomiarów składu mieszanin oddechowych w nurkowych kompleksach hiperbarycznych. PhD diss., The Polish Naval Academy.

Kłos R. 2000. *Aparaty Nurkowe z regeneracją czynnika oddechowego*. Poznań: COOPgraf. ISBN 83-909187-2-2.

Kłos R. 2002a. Metabolic simulator supports diving apparatus researches. *Sea Technology* 12, 53–56.

Kłos R. 2002b. Mathematical modelling of the breathing space ventilation for semi-closed circuit diving apparatus. *Biocybernetics and Biomedical Engineering* 22, 79–94.

Kłos R. 2003a. A mathematical model of ventilation process of a distressed/disabled submarine. *Polish Maritime Research* 4,23–25. ISSN 2083-7429.

Kłos R. 2003b. Experimental verification of a new mathematical model of ventilation of closed circuit breathing apparatus. *Polish Maritime Research* 1, 25–30.

Kłos R. 2007a. Modelowanie procesów wentylancji obiektów normo- i hiperbarycznych. Higher PhD diss., The Polish Naval Academy. PL ISSN 0860-889X nr 160A.

Kłos R. 2007b. *Mathematical modelling of the normobaric and hyperbaric facilities ventilation*. Gdynia: Polish Hyperbaric Medicine and Technology Society. ISBN 978-83-924989-0-2.

Kłos R. 2008. *Systemy podtrzymania życia na okręcie podwodnym*. Gdynia: Polish Hyperbaric Medicine and Technology Society. ISBN 978-83-924989-4-0.

Kłos R. 2009. *Wapno sodowane w zastosowaniach wojskowych*. Gdynia: Polish Hyperbaric Medicine and Technology Society. ISBN 978-83-92499889-5-7.

Kłos R. 2010. Komorowy system zachowania życia zabezpieczający górników w przypadku powstania atmosfery niezdatnej do oddychania. *Polish Hyperbaric Research* 33, 71-88.

Kłos R. 2011. *Możliwości doboru dekompresji dla aparatu nurkowego typu CRABE*. Gdynia: Polish Hyperbaric Medicine and Technology Society. ISBN 978-83-924989-4-0.

Kłos R. 2012. *Możliwości doboru ekspozycji tlenowo-nitroksowych dla aparatu nurkowego typu AMPHORA – założenia do nurkowań standardowych i eksperymentalnych*. Gdynia: Polish Hyperbaric Medicine and Technology Society. ISBN 978-83-924989-8-8.

Kłos R. 2014a. *Helioksowe nurkowania saturowane - podstawy teoretyczne do prowadzenia nurkowań i szkolenia*. Gdynia: Polish Hyperbaric Medicine and Technology Society. ISBN 978-83-938322-1-7.

Kłos R. 2014b. Inherent unsaturation. The risk of central nervous system oxygen toxicity part 1. *Polish Hyperbaric Research* 1, 37-64, DOI: HTTP://DX.DOI.ORG/10.13006/PHR

Kłos R. 2014c. The pathophysiology related to the toxic effect of oxygen. The hazard of central oxygen toxicity part 2. *Polish Hyperbaric Research* 2, 15–34. DOI: HTTP://DX.DOI.ORG/10.13006/PHR.47.2.

Kłos R. 2014d. Survival analysis. The risk of central oxygen toxicity part 3. *Polish Hyperbaric Research* 3, 33–48. Tom 48, DOI: HTTP://DX.DOI.ORG/10.13006/PHR.48.3.

Kłos R. 2014e. The hazard of central oxygen toxicity occurrence. The risk of central oxygen toxicity part 4. *Polish Hyperbaric Research* 4, 19–31. DOI: HTTP://DX.DOI.ORG/10.13006/PHR.49.2.

Kłos R. 2015. *Katalityczne utlenianie wodoru na okręcie podwodnym*. Gdynia: Polish Hyperbaric Medicine and Technology Society. ISBN 978-83-938-322-3-1.

Kłos R. 2015b. Measurement system reliability assessment. *Polish Hyperbaric Research*. 2, 31–46. DOI: 10.1515/phr-2015-0009.

Kłos R. 2016. *System trymiksowej dekompresji dla aparatu nurkowego typu CRABE*. Gdynia: Polish Hyperbaric Medicine and Technology Society. ISBN 978-83-938322-5-5.

Kłos R. 2017. Pollutions of the hyperbaric breathing atmosphere. *Scientific Journal of Polish Naval Academy.* 208, 31–44. DOI: 10.5604/0860889X.1237621

Kłos R. 2019. Modelling of the normobaric and hyperbaric facilities ventilation. *International Journal of Mechanical Engineering and Applications* 7, 26–33. DOI: 10.11648/j.ijmea.20190701.14.

OTHER CITED PUBLICATIONS

Birch K., MacLaren D., & George K. 2009. *Fizjologia sportu-krótkie wykłady.* Warszawa: Wydawnictwo Naukowe PWN. ISBN 978-83-01-15460-8.

Dąbrowski J. 2000. *O problemie redukcji wymiarów.* Kraków: Polskie Towarzystwo Inżynierii Rolniczej. 83-905219-4-6.

Dobosz M. 2001. *Wspomagana komputerowo statystyczna analiza wyników badań.* Warszawa: Akademicka Oficyna Wydawnicza EXIT. ISBN 83-87674-75-3.

Dräger Safety AG & Co. KGaA. 2003. *Multi-Gas Monitor Multiwarn II Technical Handbook.* Lübeck: Dräger Safety AG & Co.

Frånberg O. 2015. Oxygen content in semi-closed rebreathing apparatuses for underwater use: Measurements and modeling. PhD diss., Stockholm School of Technology and Health. ISSN 1653-3836.

Goliński JA, & Troskalański AT. 1979. *Strumienice.* Warszawa: WNT.

Haux G. 1982. *Subsea manned engineering.* London: Bailliére Tindall. ISBN 0-7020-0749-8.

Kłos I., & Kłos R. 2004. *Polish Soda Lime in military applications.* Oświęcim: Chemical Company DWORY S.A. ISBN 83-920272-0-5

Kyriazi N. 1986. *Development of an automated breathing and metabolic simulator.* Pittsburgh: Department of the Interior: USA Bureau of Mines, No NC6414. 614.894

Loncar M. 1992. Breathing simulator simulates human oxygen intake. *Offshore* 10, 82.

Minilab Mod.No. 92-25. 1992. *Operators manual.* Husøysund: Ottestad Breathing Systems.

Mittleman J. 1989. Computer modeling of underwater breathing systems. In. *Physiological and human engineering aspects of underwater breathing apparatus*, ed. DE Warkander & CEG Lundgren, 223–270. Bethesda: Undersea and Hyperbaric Medical Society.

NATO Standardization Agreement STANAG 1320 UD. 2005. *Minimum requirements for atmospheric monitoring equipment located in submarines with escape capability.* Brussels: NATO Standardization Office. NSO(NAVAL)1449(2014)SMER/1320.

NEDU. 1994. *U.S. Navy unmanned test methods and performance goals for underwater breathing apparatus.* Panama City: US Navy Experimental Diving Unit. Report No 01-94.

PN-EN 250. 2014. *Respiratory equipment – Open-circuit self-contained compressed air diving apparatus – Requirements, testing and marking.* Brussels: CEN-CENELEC Management Centre.

Przylipiak M., & Torbus J. 1981. *Sprzęt i prace nurkowe-poradnik.* Warszawa: Wydawnictwo Ministerstwa Obrony Narodowej. ISBN 83-11-06590-X.

Reimers SD., & Hansen OR. 1972. *Environmental control for hiperbaric applications.* Panama City: US Navy Experimental Diving Unit. NEDU Rep. 25-72.

Segedal K., & Morrison JB. 1985. *Acceptable criteria and unmanned test procedures for underwater breathing apparatus.* Bergen: Norwegian Underwater Technical Centre, 1985. Report NUTEC 17-85.

Williams S. 1975. Engineering principles of underwater breathing apparatus. In *The physiology and Medicine of diving*, ed. PB Bennett & DH Elliott, 34-46. London: Baillière Tindall.

US Navy Diving Manual. 1980. Carson California: Best Publishing Co. NAVSEA 0994-LP-001-9010.

US Navy Diving Manual. 2016. The Direction of Commander: Naval Sea Systems Command. SS521-AG-PRO-010 0910-LP-115-1921.

Appendix

Some Principles Regarding Participation of Polish Navy Personnel in Experiments

There are different theoretical bases of mathematical models. If there are no adverse experiment-based reasons, the choice of the specific mathematical model is based on the subject's preferences. In order to obtain hyperbaric data, it is necessary to carry out experimental diving. It is a very difficult, expensive, and time-consuming method, and the results are sometimes controversial and ambiguous. However, it is the only way of getting reliable results.

Sometimes there is a tendency to abandon deterministic models[1] in favor of statistical models. If we have only a small volume of experimental data, we must take into account that reliable statistical models cannot be established.[2]

The detailed description of each stage of experimental investigations was previously presented in this book.[3]

Based on the experimental and theoretical knowledge, new mathematical models are developed during the experimental phase of research. The next stage of research includes testing the models. Within the framework of this research stage, the assumptions underlying the research should be made in compliance with the Declaration of Helsinki. Some of the principles observed by the navy are presented below.

THE BASIC PRINCIPLES

1. Medical experiments must be consistent with the scientific and ethical principles that motivate medical research. They should be based on laboratory tests and medical experiments with animals or other scientific circumstances.
2. Medical experiments should be performed only by the qualified scientific personnel under the supervision of a qualified medical supervisor.[4]
3. Medical experiments cannot legitimately be carried out unless the importance of the objective circumstances is proportional to the risk undertaken.
4. Each scientific project of medical research should be preceded by careful evaluation of the risk being incurred with respect to the possible expected results[5] for the subject or humankind.
5. When medicines or experimental procedures clearly cause the subject harm, the medical officer who is responsible for the medical experiments should undertake appropriate precautions.

6. In order to obtain new medical knowledge, the therapist can apply the combination of the medical experiments with the standard method of the therapy. This is allowed only during medical treatment.

THE INDEPENDENT MEDICAL EXPERIMENTS

1. During involving human subjects, the doctor is obliged to take measures to protect the health and life of the subject who participates in the experiments.
2. The doctor has to explain to the subject the character, objective(s), and risk(s) of the medical experiments.
3. Medical experiments on human beings cannot be undertaken without the subjects' agreement obtained after they have been informed about the purpose, character, and risks involved in the experiments. When the subject is legally unable to give their consent, the agreement should be received from the person responsible for their care.
4. The subject should be mentally, physically, and legally capable of making his/her own decision regarding his/her participation in the experiments.
5. The agreement should be obligatory and in writing. Independent of the written agreement, the scientist is always responsible for the experimental results.
6. The researcher must respect the rights of each subject to maintain his/her personal integrity, particularly when the subject remains officially dependent on him.
7. Usually, during medical experiments, the subject or the person in his/her custody has the right to withdraw from experiments. The researcher or research team should then discontinue the experiments.

In Poland it is necessary to obtain permission to conduct medical experiments, issued by the Ethics Research Board. The laboratory experiments include tests in a hyperbaric chamber, in a water tank, as well as in a natural body of water. The experiments should be performed as long as the results are approved by the medical staff in attendance. The results obtained enable the beginning of the implementation stage. The implementation stage can be discontinued at any moment if any problems arise. Changes in the program may entail changes in the model. If an amendment in the guidelines for the experiments is needed, it should again be approved by the Ethics Research Board. The number of laboratory tests enabling the implementation stage is dependent on the tested model. In the case of earlier model testing and insignificant changes in its construction, a fewer number of tests is required. In order to reach the proper level of confidence in the course of testing the new model, the number of tests must be sufficient to achieve the required results.

The decision to complete the laboratory research and to implement the results is taken by the relevant military authority. For the Polish Armed Forces, the Armament Policy Department, the former Development and Implementation Department, is the relevant military authority. In Poland, Chief of the Search and Rescue Department submits the motion to the Commander of the Polish Navy, who is authorized to take appropriate decisions, and a special Board is established. It consists of a safety officer, a medical officer, a representative of the headquarters, and some specialists.

Investigations are carried out in compliance with the supplementary instruction to the Diving Rules. The safety rules are very strict. The research group includes a medical and research team as well. Additional measuring instruments are used for data acquisition in the course of diving experiments, etc. If there are no problems observed at the implementation phase, a decision on implementation can be made. The research process is continued. The results of diving are periodically analyzed by specialists during normal operation.

If there occur any problems with operating the diving system, repeatable investigations within the framework of the introductory phase should be carried out. If the diving procedure is canceled or modernized, the process will start from the beginning. The method seems to be complicated, but it can be employed in military conditions.

NOTES

1 Cause-effect mathematical model.
2 Empirical and semi-empirical models are the most common ones.
3 The scheme for the procedure is presented in Figure 7.1.
4 Also known as controller.
5 Specifically the advantages.

Index

A

C

D

E

G

H

M